COMING OF AGE WITH ELEPHANTS

COMING OF AGE
WITH ELEPHANTS
A Memoir

BY JOYCE POOLE

HYPERION

NEW YORK

Library of Congress Cataloging-in-Publication Data

Poole, Joyce, 1956–
Coming of age with elephants : a memoir / Joyce Poole.
p.   cm.
ISBN 0-7868-6095-2
1. Poole, Joyce, 1956–   2. Ethologists—Africa—Biography.
3. Women ethologists—Africa—Biography.   4. African elephant—Behavior—Kenya—Amboseli National Park.   5. Amboseli National Park (Kenya)   I. Title.
QL31.P68A3   1996
591.5'092—dc20
[B]   95-40153
CIP

Book design by Holly McNeely

FIRST EDITION

1   3   5   7   9   10   8   6   4   2

To Selengei,
the delight of my life,
to the wonderful man who gave her to me,
and
to Richard,
whose courage, generosity,
and support made everything possible.

Contents

Acknowledgments *ix*

PROLOGUE: A Flight to Lake Natron *1*

PART I   Early Days Among the Elephants 1962–1978

   1. An African Childhood *7*

   2. A Return to Africa *16*

   3. The Elephant Camp *22*

   4. The Five Males *28*

   5. Naming an Elephant *36*

   6. The Green Penis Syndrome *43*

   7. The Lasting Sorrow *50*

PART II   Field Biologist 1979–1984

   8. Designing My Study *57*

   9. Male Watching *65*

   10. Dangerous Encounters *74*

CONTENTS

11. My Amboseli Home *81*

12. Births and Deaths *92*

13. A Sojourn in Cambridge *100*

14. A Maasai Named Meloimyiet *108*

PART III   Reflections on Elephants

15. Elephant ESP *119*

16. Trunks, Tusks, and Tool Use *136*

17. Elephant Thinking *142*

18. The Games Elephants Play *149*

19. An Elephant's Empathy *159*

20. Amboseli Seasons *166*

PART IV   A Removal of Veils 1984–1989

21. The Incident on the Ngongs *177*

22. Fear and Loathing in Amboseli *182*

23. The Elephants' Graveyard *192*

24. Dire Predictions *204*

25. A Poacher's Story *215*

26. Tsavo Survival *220*

27. Burnt-to-Blue Ivory *228*

PART V   A New Life 1990–1993

28. A Voice for Elephants *237*

29. A Cache of Ivory *243*

30. The Elephant Menace *249*

31. A View Over the Rift *257*

32. The Birth of Selengei *267*

33. A Family Reunion *273*

EPILOGUE *276*

INDEX *279*

Acknowledgments

YEARS HAVE PASSED SINCE I earned the title *Mama Ndovu*, Mama Elephant, and much has taken place in my life, and in the lives of the elephants I have been so privileged to know. I am indebted to many who have helped me along the way, and though I mention only a few by name, I hope the others will recognize themselves in the words below.

I remember especially my mother and my father who gave me the Nature and Nurture that have made my life what it is; my siblings, Bobby and Ginny, who inherited and shared them with me; and my young daughter, Selengei, to whom I will endeavor to pass on the very best of them. I remember, too, my friends, who have given me so many reasons to laugh and to love, and who have supported me through the difficult times. I give my special thanks to: Richard and Meave Leakey and their daughters, Louise and Samira; Iain and Oria Douglas-Hamilton and their daughters, Saba and Dudu; Cynthia Moss, Barbara Tyack, and Cynthia Jensen who have, during different periods, each provided me with "family."

ACKNOWLEDGMENTS

The friends and colleagues with whom I shared the Amboseli elephants will always have a special place in my heart and I thank Norah Njiraini, Soila Sayialel, Keith Lindsay, Phyllis Lee, Katy Payne, Kadzo Kangwana, and, in particular, Cynthia Moss, who introduced the elephants to me. I am indebted to the camp staff, Masaku Sila, Peter Ngande, and Vincent Wambua, for their tireless assistance. I am grateful to my colleagues, particularly those I worked with in Amboseli, and at Cambridge, Princeton, and Cornell Universities, who shared ideas and knowledge; my teachers, professors, and supervisors who inspired me to ask the question, *why?*; each of the institutions that have supported my research over the years and, in particular, the African Wildlife Foundation, which provided the Amboseli Elephant Project and me with a home base. I thank the Kenya Government, particularly the Wildlife Department, for allowing me to work in Amboseli National Park, and the Amboseli wardens and rangers with whom I shared a deep commitment to the elephants.

During the years that this story covers, Africa's elephants went through some dire times, and I extend my gratitude to people everywhere who, in their own way, fought for the protection of the elephants. Close to home this includes the rangers and wardens who, despite the hardships, endeavored to make a difference; the organizations and individuals who didn't sit on the fence; the producers, journalists, filmmakers, and photographers who tried to tell the real story.

I feel privileged to have worked with an extraordinary group of individuals at Kenya Wildlife Service, and I thank, in particular, each one of my "elephant team" for their enthusiasm and dedication to the task. Many projects undertaken by the Elephant Program would have been impossible without the financial and material support, and the guidance of numerous institutions and individuals, and I thank each one of them for their generosity, patience, and understanding. I am grateful to the heads of other departments and programs at KWS for including me as "one-of-us" not "one-of-them," and of course to the former director,

Richard Leakey, for his faith in me. I developed some deep and loyal friendships at KWS which have lasted through the difficult and often traumatic times that followed, and I thank in particular Carole Mwai Wainaina and Tahreni Bwana for their friendship and support.

I am indebted to my architect friend, Francisco Siravo, who helped translate my ideas for a house into a design that, I hope, will not slide down the 25-degree Rift Valley escarpment; the engineers who gave their input; the masons, carpenters, plumbers, and electricians who struggled to make my dream a reality. Most of all I thank, once again, my friend and neighbor, Richard Leakey, who liked to refer to himself as my "contractor," for the promise he made and kept to help me build a home of my own, and to Meave, Louise, and Samira for their extraordinary hospitality, their friendship, and their tolerance.

I thank those who encouraged me to tell my story, those who read through the manuscript and gave me advice on how it could be improved, and those who donated their photographs. In particular, I am grateful to my editor at Hyperion, Rick Kot, who won my confidence through his adept handling of some sensitive issues.

I am deeply grateful to all of my friends who played a role, however small, in bringing Selengei into my life, and to those whose friendship will continue to make her life warm and rich. I owe special thanks to Pauline Wathoni and Martha Khayanje, who looked after Selengei for long hours while I wrote, and to Selengei herself, for her lasting tolerance.

Acknowledgments would not be complete without saying that I will be forever indebted to the elephants for just being themselves, for giving me a life of meaning and days of joy.

Author's Note

Numerous elephant books have been written, and many elephant tales have been told. I have chosen to recount a personal tale of my experiences as a young woman living in Africa in the company of elephants. I might have written a scientific account of my research, but I wanted to share, without so many restraints, my love for elephants and for Africa with a wider audience.

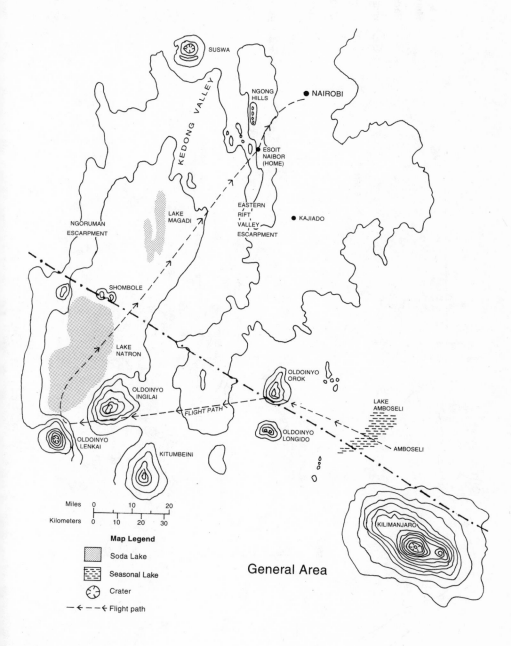

General Area

SUSWA

KEDONG VALLEY

NGONG HILLS

● NAIROBI

ESOIT NAIBOR (HOME)

EASTERN RIFT VALLEY ESCARPMENT

● KAJIADO

NGORUMAN ESCARPMENT

LAKE MAGADI

SHOMBOLE

LAKE NATRON

OLDOINYO INGILAI

FLIGHT PATH

OLDOINYO OROK

LAKE AMBOSELI

AMBOSELI

OLDOINYO LONGIDO

OLDOINYO LENKAI

KITUMBEINI

KILIMANJARO

Miles   0   10   20
Kilometers   0   10   20   30

Map Legend

Soda Lake

Seasonal Lake

Crater

—←——←— Flight path

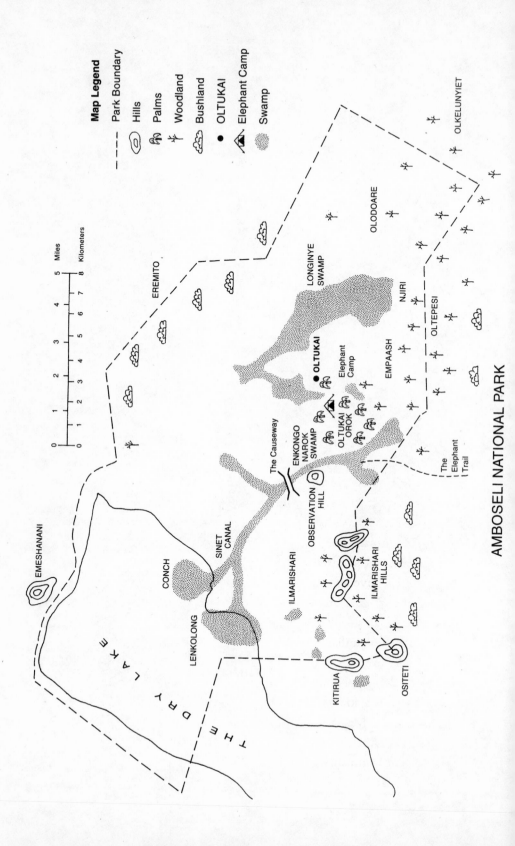

AMBOSELI NATIONAL PARK

Prologue

A FLIGHT TO LAKE NATRON
FEBRUARY 1991

RICHARD LEAKEY AND I left Amboseli National Park in southern Kenya in the early-morning light, flying due west toward Lake Natron. From the air the plains looked cool and damp, the colors soft and restful. The ground below us had a peaceful appearance as if the land and its inhabitants had still not woken from the long night. The usually brilliant volcanic ash, which during the heat of the day swirled across the plains in wild, furious dust devils, was now a soft mauve and lay still and quiet. Lone male wildebeests lay dotted about, guarding their territories, and the elephants stood around in small sleepy groups. As the aircraft pulled away from the earth I felt the familiar contradiction stir in my chest—a deep longing for a return to the peace and tranquility of the bush, and the relief of leaving it behind.

Amboseli is a strange land, at once barren and yet so full of life. *Empusel,* as the Maasai call it, where the dust flies up from the lake. Long ago a river flowed through here, from west to east, but the eruptions of Kilimanjaro had blocked the river's course and creating

1

a lake with an inlet but no outlet. With time the lake dried up, leaving crusty deposits of salt across the plains. The original lakeshore can still be seen where the soil changes to a soft rust color on the ridge in the north, and where a band of *Acacia tortilis* grows in the south. It is here, among these umbrella-shaped trees, that the land begins to rise toward Kilimanjaro. The soil of the Amboseli basin itself is a brilliant ivory-colored powder, a mixture of salts deposited over thousands of years by the river and volcanic ash from the eruptions of the numerous small cones and craters that punctuate the flat landscape.

The main feature of the land is snowcapped Kilimanjaro, *Oldoinyo Oibor,* the White Mountain, rising 4,500 meters above the lake basin. Kilimanjaro not only dominates the land physically but governs the very pulse of life, for though Amboseli is categorized as semiarid savannah, bordering on desert, the mountain's porous volcanic rock and underground aquifers channel its rainfall and melting snow to Amboseli, where it bubbles out in springs, creating a series of permanent swamps. These swamps are the only source of food and water during the long dry season, and have provided a home for a myriad of domestic and wild animals for thousands of years.

Our Cessna 206 followed the emerald-green line of papyrus along *Enkongo Narok* swamp to where it spills onto the dry lake, creating a sphere of luxuriant vegetation. From the air the dry lake bed was crisscrossed with an intricate pattern of a thousand trails, all converging on the lush green of the swamp. It was the height of the dry season, and Maasai herdsmen were already walking their cattle toward the only remaining food and water. The cattle made their way slowly, a long single file in the colors of pottery. *Lenkolong,* as the area is called, the place where the sun beats down all day. The scene reminded me of a Maasai friend who here, as a young boy, had been forced to drink the urine of his cattle to quench his thirst.

We left Amboseli behind and continued for half an hour westward over an expanse of desolate land, forgotten and cut off from the outside world. The earth suddenly softened, and rounded hills fell gradually into valleys where green-leafed fig trees followed the dry riverbeds. We flew over the shoulder of *Oldoinyo Gelai,* where stunted acacias dotted the brown hillsides. The grass was long, and, bent over by the wind, it shimmered golden in the morning light. Suddenly my eyes caught a flash of radiant blue and pink ahead. This was what Richard had brought me to see. A place he had explored many years ago, during his paleontological days.

To our left *Oldoinyo Lenkai,* the Mountain of God, an active volcano, stood as a tall, steel-blue monument to the Great Rift Valley. Streaked with deeper blue ravines of eroded ash, it guarded the southern entrance to Lake Natron. The Rift Valley runs from Syria to Malawi, and is one of the few geological features of the earth that can be seen from the moon. Lake Natron itself can be observed from hundreds of kilometers into space. Richard dipped the right wing of the aircraft so that I could see the alkaline lake: octagonal cells of deep, lurid pink swirls of encrusted soda, curving round and round; circles within circles of dazzling pink. Each cell of swirls was outlined and separated from the next by a frothy-looking crust of white soda. Richard banked the plane and we circled again. Thousands of brilliant pink flamingos took to their wings below us. I was mesmerized by the colors, the patterns, the wing beats, the changing horizon, the sky, the lake, the swirls drawing me closer and closer. As if reading my mind, Richard said, "If we fell in, we would never reach the shore. We would be burnt alive within minutes." My eyes were drawn again to the intense pinks, imagining the crunching of soda, the burning sensation; it would be one way to die. As Richard pulled the plane out of the curve, up and away from certain death, I looked at the brilliant blue of the sky, the steep escarpment that marked the western Rift Valley wall and then back down the length of the lake to *Lenkai.* Over the drone of the engine I heard him say *"Shombole,"* and I turned to look at *Lenkai*'s soft, rounded sister guarding the north end of the lake, and the Kenya–Tanzania border.

The pinks, the blues, the unbearable harshness of the landscape was what made it so exquisitely beautiful, I thought. It was this variety, the striking series of contrasts that is Africa, that had kept me staying on. In the soft morning light the shapes and colors had a mystical, ephemeral quality. An aching sensation swept over me that was as strongly physical as it was emotional. At any moment the magic would be gone. I felt the beauty tearing at my heart, shaking my senses. I stared hard out of my window blinking back the tears. Any minute now Richard would ask what was wrong, and what could I say? If it was hard enough to understand myself, how could I ever explain it to anyone else? Would he understand if I told him that this was the way the elephants standing in the moonlight, the wind on the plains, and the last light on the snows of Kilimanjaro always used to make me feel, but after everything that had happened, I had feared I would never be able to feel the beauty of Africa again? Or perhaps, like the Africans, I could simply say "*Ikapotea na nimeitafuta kwa muda mrefu,*" that somehow it had gotten lost, and I had been searching for it for so long, and now that I had finally found it I couldn't bear the thought of ever losing it again.

PART ONE

EARLY DAYS AMONG THE ELEPHANTS 1962–1978

CHAPTER ⟫1

An African Childhood

NIMEKULA ASALI UDOGONI, UTAMU UNGALI GEGONI.
I ATE HONEY IN MY CHILDHOOD, AND ITS
SWEETNESS IS STILL IN MY TOOTH.
—A Swahili Proverb

AFRICA IS THE WAY the wind feels against my face as it blows across the dry Amboseli plains, or the damp, earthy smell of the first rains. It is watching the last rays of the sun cast a touch of pink on the western snows of Kilimanjaro, being woken by the full moon shining in through the window of my tent, or waking to see the layered branches of the acacias silhouetted against the first light of dawn. Africa is the glint of an elephant's tusks in the moonlight, silent shadows cast upon silver soils, or the sound, in the darkness, of an elephant's footsteps on a crusty saline pan.

Africa is the heady sensation of singing and dancing in the midday heat and dust with the Maasai warriors, or the calm of drinking tea with the women in the smoky darkness of their homes. It is entering the women's homes, relying on smell and touch to guide myself along the rough, windowless walls of mud and dung, or sitting on a hard bed of branches covered with cowhide worn smooth from years of use, pressed against the warm bodies of other women, the smell of sour milk, urine, smoke, and sweet manure invading my senses.

7

Africa is the intense awareness I feel when walking through the bush alone, alert for the sounds that mean danger, or of driving across the sands of the Selengei, conscious that my wheels may begin to spin. It is reaching out to hold Vladimir's tusk or knowing to within a meter how close I can approach Bad Bull and still escape when he attacks, confident that I can read the moods of wild elephants better than perhaps anyone in the world.

For me, the magic of Africa is the freedom and the challenge of life on the edge—the very certainty of uncertainty. There is a poignance to life in Africa, a knowledge that you will be touched deeply by it, and yet, in a sense, you will never truly belong.

I was born in Europe of American parents; I am, by nationality, American, but that heritage seems to me, like the place of my birth, an accident of circumstance. I moved to Africa when I was a young child; in my heart I am African, white African. My thinking and my behavior are, I believe, a blend of several cultures. In some ways my outlook is more African than that of many of my close African friends; in other ways, I am told, my Americanness stands out. My sense of outrage, for example, when life is not free and fair, though I know perfectly well that I should not expect it to be. I feel the influence, too, of a third "culture," another way of looking at the world: through the eyes and ears of the elephants. Other elephant people have said they suspect I believe I am, in fact, an elephant. In some ways, perhaps, they are right. Like Africa, the elephants take hold of your spirit.

It is strange, now, to think back to when much of Amboseli National Park was for me a great unknown; when there were places I had never been to, where beyond certain limits there was only a vast wilderness to explore. Twenty years ago, at the beginning of my study, the Amboseli elephants for me had no names, individual trees and bushes had not yet become familiar shapes on the horizon from which to distinguish elephants, and *Oltukai Orok,* the dark place of the palms, was an intricate maze of cul-de-sacs, an area in which I frequently got lost.

Longer ago Amboseli was, for me, the setting of family safaris. I have vivid memories of squinting my eyes against the dazzling brilliance of the midday sun, trying to count zebras through the shimmering heat haze, as we chased mirages endlessly across the dry lake bed: black and white stripes quivering and melting into space. Many memories of Amboseli from my early childhood revolve around old photographs. One, in a family album, captioned *Amboseli 1967*, shows my brother and me posing nervously in front of Odinga, a famous elephant. Several years later Odinga was shot by an American hunter who waited patiently for him to cross the boundary of the reserve where he passed from magnificence to a pair of trophies. Odinga, the elephant who was given treats, hand to trunk, from the lodge kitchen.

I remember even longer ago, as a child of seven on safari to Kenya from Malawi, staring up at a huge bull elephant and asking my father what would happen if he charged our Land Rover. I can still feel the rush of adrenaline as the bull came for us, his great flapping ears, the cloud of dust, and my father's answer—"We will be squashed down to the size of a pea pod"—catching in my throat. A photograph of the elephant in midcharge, now faded with time and spoiled by a spidery network of violet mildew, lies stored among the many boxes of wildlife slides taken by my father.

Another photograph shows a dark line of Maasai warriors clad in leather, painted with ocher, decorated with beads and stuffed birds, standing in a landscape of brilliant volcanic ash. Slightly to the left of center my mother poses awkwardly in a neat dress, her hair carefully curled, holding my golden-haired sister, Virginia, age one. By her side, my red-headed brother, Bobby, age four, squints impatiently at the camera. Dressed for the occasion in a pretty blue-and-white dress, I stare up at the warriors behind me. My father, always the man behind the camera, is absent. *On safari, Amboseli, 1963*. Images, capturing a moment in time.

My father put together three large albums, starting in Malawi in 1962 and ending in Kenya in 1969; they contain a selection of his best photographs. They were a portrait of a family and a way of life. It had seemed to me, even then, that we were not like most other American families. In the early 1960s few Americans had reason to travel to Africa, and when we left Litchfield, Connecticut, to move to Africa for the first time in 1962, the news of our departure was big enough to deserve a photograph and story in the local newspaper. In it Bobby and I stand together at the base of the stairs, holding a red-and-white-checkered bag of toys between us, waiting to board TWA for the long journey ahead. The Peace Corps had just been established, and my father had been named director of the program in Malawi, then pre-independence Nyasaland. His selection was based on the fact that he was a graduate of Yale University, he had taught African history at a prestigious boys' boarding school, and he had recently hitchhiked around the African continent. A knapsack on his back, binoculars at hand for birdwatching, he had been mistaken for a spy and thrown into jail in the Congo. By American standards, he was considered an expert on African affairs and seemed a perfect candidate for the ideals of the new Peace Corps.

Our arrival in Malawi was less than perfect. There was no house available for us, and we had to stay in a hotel in the town of Limbe for six weeks. While my father settled into his new job, my mother was left to care for three young children. I was then six years old, my brother three, and my sister just eight weeks. It is odd how certain memories from early childhood stand out so clearly, while others are lost forever. I still carry images of events that took place during those first few months in Malawi that, in their clarity, could have taken place yesterday. My brother spent the first few days informing the hotel staff that they couldn't come with us to Africa because they would be eaten by lions, and I marveled at the whole bees in the honey provided at breakfast. My mother became very distressed when she discovered a large putzi fly larva embedded in my infant sister's back, which had to be squeezed out.

My mother had believed that joining the Peace Corps meant that we would live like other people, in a simple home possibly with a thatched roof. But when we left the hotel we moved into a large, imposing white house on a hill surrounded by a grove of trees. It seemed like a mansion to me, with a sweeping stone staircase leading up through a beautiful garden to the main entrance. Initially we had no furniture, no curtains, and our voices echoed through the halls. Bushbabies screamed in the night and stared in through the bare windows with their huge eyes.

My brother and I led outdoor lives. We were fascinated with the various creatures we discovered in our garden, but our parents reminded us of the numerous diseases we could contract. A small pond on the grounds filled with lily pads, frogs, and goldfish captured our attention, but we were told not to put our hands in the water for the fear that we might catch bilharzia. We were not allowed to walk barefoot because, we were informed, microscopic hookworms would burrow into our feet, though a friend of mine who never wore shoes told me that she simply jumped over them. And we had a dreadful Sunday morning malaria prophylactic-taking ritual that, despite the honey, peanut butter, and jam, took me hours to complete. Notwithstanding the great care our parents took, my brother came down with a severe case of malaria within the first few months of our arrival. Bobby had a high fever with hallucinations, and my parents stayed by his bedside through the night removing the "hundreds of lizards and snakes" that he believed to be on the mosquito netting.

I had a young school friend by the name of Heather and near to her house was a large rock, or *kopje,* which we liked to explore. On top of the *kopje* was a vertical pipe stuck into the rock with cement. One of the house servants had warned us that a deadly black mamba came out of the pipe every day at four o'clock, and, followed by her many babies, she slithered around the rock seventeen times before going back down the pipe. Heather and I dared each other to look down the pipe, though we knew that if bitten by the snake we would die within minutes.

While we never saw the mamba, I was bitten on numerous occasions by vervet monkeys while staying at a holiday resort on the shores of what was then known as Lake Nyasa. The proprietor was an eccentric blond woman who had habituated numerous wild animals, including several that were permitted to sleep on her bed. A number of aggressive monkeys made a game out of terrorizing small children, and one pulled down my bathing suit and bit me hard on my bottom.

Once upon returning from a lake holiday we had a car accident, and I hold strong memories from that dark night: cattle running across the road in the headlights; the screeching of car brakes; the swerve of the Land Rover; my sister falling out of my mother's arms onto the floor. The cold night air contrasted sharply with the warmth of the car headlights and the last bit of life ebbing from one cow. My father wanted me to stay in the car, but I remember seeing its glassy eyes wide and staring and the blood flowing from its warm body as the owners slit its throat.

Afterward we stopped at a nearby settler's home while my father spoke to the police. As my mother carried on polite conversation, I studied the objects in the sitting room, struggling to stay awake. In typical colonial fashion, it was furnished with large overstuffed armchairs in a floral print worn and faded with time. Long silken tassels dangled from the lampshades, stained brown in places from the heat of the bulbs. Paintings of tame European landscapes decorated the walls. Then I saw the elephant foot, now a container stuffed with old newspapers and magazines. I had never seen a wild elephant before, and I knew nothing about them, but the idea of cutting the foot off an animal and using it as an ornament was distressing to me. I studied the skin and the toenails. There was no blood now, no pain, but like the cow, this elephant must once have been free, fleeing for its life.

After eighteen months in Malawi, during which we took our safari to Kenya and Amboseli, my father was promoted and we moved to Washington, D.C., where he was to head the Peace Corps Africa Program. At school in Nyasaland, I had been the only American; when we returned to America, I was immediately singled out as the child from Africa. Often I was asked whether I had lived in a mud hut—such was the knowledge of Africa gleaned primarily from Hollywood films and photographs in *National Geographic*. My grandfather always introduced me with great pride: "This is Joyce, my African grandchild." At this people raised their eyebrows and then seemed at a loss for words. It was as if they were thinking that since Africa was a strange and mysterious sort of place, I must, by extension, be a strange and unusual sort of a child.

It was a mere nine months before we were once again packing up our belongings and returning to Africa. My father, a naturalist at heart, found the hustle and bustle of city life, the long hours at a desk, not to his liking and requested another posting in Africa. So in 1965 we found ourselves on our way to Kenya.

In the 1960s most white children in Kenya went to schools that were still very much racially segregated, even if not necessarily by intent. I attended Hospital Hill Primary, Kenya's first international school, which made its progressive reputation by being "one-third African, one-third Asian, and one-third European," and then later Kenya Girls High School, which was, by then, 90 percent African. So strong an impression did my school experience make that I still remember the names of each of the children in my class. It is interesting, today, how many influential and successful Kenyans I meet who went through the same school system, and I am grateful to my parents for exposing me at a young age to people of different backgrounds.

My father taught me to drive a car when I was ten, while my mother tried to ensure that I was engaged in all manner of after-school lessons. During the four years we spent in Kenya, I was instructed in ballet, tennis, swimming, horseback riding, art, singing, recorder, and guitar. In retrospect, I cannot imagine how she or I fit it all in. Since my brother and sister were also involved, I know that she would say "Wearily." Eventually I rebelled. I particularly disliked Pony Club, which I found far too stuffy, and after my pony, Gilbert Golly, bit an examiner, causing me to fail one of the tests, I flatly refused to waste any more Saturdays.

We lived on five acres of land that, beyond the horse stables and the banana and avocado trees, sloped down through the bush to a river. By the river were some caves that were rumored to have been hideouts for members of the Mau Mau during the 1950s Emergency and, during the 1960s, for the so-called Panga Gangs. *Panga* is the Swahili word for "machete," and occasionally houses in the vicinity were raided and robbed by these machete-wielding gangs. The valley and the river had to be crossed when I went to visit my school friend Amanda, and, for me, the journey held great excitement.

My father built us a fabulous tree house that could sleep five on camp beds. The tree house had a trapdoor as well as a ladder on a pulley system, and it became the scene of numerous neighborhood battles between different factions, usually boys versus girls. We had more pets than my mother cares to remember, since she was the one who usually ended up caring for them. They included two horses, two dogs, a cat, four bushbabies, scores of rabbits and guinea pigs, chickens, one monitor lizard, three hedgehogs, six snakes, and, of course, mice and frogs to feed the snakes. I often carried my snake, Edward-the-Egg-Eater, with me to the breakfast table. My mother found another of our pets, a semipoisonous white-lipped snake, in bed with her one night. One of our horses, Dobbie, had been used in the movie filmed in Amboseli, *Where No Vultures Fly*. He was a very clever horse and had learned to open his own stable, and Gilbert Golly's, and then to let themselves out of the front gate, where they would take off down the road. Once he managed to get into the kitchen, where my mother found him eating out of the bread box.

Whenever my father could get away from the office, we spent holidays either camping along Kenya's beautiful coastline or in its extraordinary wildlife areas, and Amboseli was one of the special places that we visited. Ours was a life of great freedom and adventure, and it seemed to me that our years in Africa gave us a closeness that other families didn't have. Perhaps all the moving and the changing of schools limited the development of long-term friendships with other children, and for that reason family bonds seemed the most reliable and strong.

I was thirteen years old when we left Kenya and returned to the United States again. It was 1969. Americans were preoccupied with the Generation Gap, the Vietnam War, Woodstock, making peace and love, and smoking dope. I did not understand the slang, the attitudes, or the rebellion. It took time for me to learn the social codes of American teenage dress and behavior, which were set by the media and intense peer pressure. I felt as if I were under assault, and I longed for the freedom of expression that I associated with Africa. With time I adjusted, but I never lost my deep desire to be in Africa. It was a dream that I shared with my father: I would return to Africa, somehow, someday.

I was in my first year of college in 1975 when the family was presented with an exciting opportunity: My father had been offered a chance to work in the field of wildlife conservation, heading the Nairobi office of what was then the African Wildlife Leadership Foundation. I was studying biology, the subject I had chosen at the age of eleven after hearing Jane Goodall speak at the National Museums of Kenya on her fascinating study of chimpanzees. After discussions with my parents, it was agreed that I could take a year off and return with them to Kenya, as long as I applied myself to a worthwhile project. At that time East Africa was a mecca for young biologists engaged in field research in the behavior of African mammals, and I longed to participate.

In August 1975, full of excitement and anticipation, we boarded the plane to Nairobi. Only my mother was reluctant to return. Silent tears streamed down her cheeks, as if she had some premonition that this was where it would all end, that in Africa the dreams of a lifetime would come to lie in shattered ruins.

CHAPTER ⟩2

A Return to Africa

I PRAYED THAT ALL THIS LOVELY WILDERNESS WOULD NOT FORGET ME,
AND THE ANSWER WAS ALWAYS: "IT IS NOT I WHO WILL FORGET,
IT IS FOR YOU TO FIND YOUR WAY BACK."
—Vivienne de Watteville, *Speak to the Earth,* 1935

I RETURNED TO KENYA with high hopes that I would be able to study one of Africa's large mammals, and elephants were top on my list. I thought I already knew quite a bit about them. I had, after all, grown up in Africa, and most of my childhood holidays had been spent on safari in the national parks. One of my earliest memories was of being charged by an old bull elephant, and since then I had spent many sleepless nights on safari in the safety of the car, waiting for the elephants to go away so that we could return to our tents. These and other childhood experiences had left me with an impression that elephants were very large, unpredictable, aggressive, and dangerous. I knew, from common knowledge, that it was not advisable to get too close to a group of them and that, if they began to flap their ears, it was best to move away, as this was a clear signal that they were annoyed. The car engine should always be left running, because elephants could charge without warning.

More recently I had read Iain and Oria Douglas-Hamilton's book

16

Among the Elephants and had learned about their multifaceted lives: their complex social structure, their intelligence, their apparent understanding of death. I learned from the Douglas-Hamiltons' study of the Manyara elephants that females and their calves lived in stable groups called family units and that apparently related families combined to form larger kin groups. The males, it seemed, were less sociable and lived separately, either on their own or in small bachelor groups. Conventional wisdom was that the older bulls were reproductively inactive and spent most of their time in all-male groups. Elephants still charged humans unprovoked, as the Douglas-Hamiltons recounted, but in general there seemed more rhyme and reason to their behavior than I had previously been led to believe.

My boyfriend, Paul Klingenstein, had also taken time off from Harvard University to join me that year, and I was secretly disappointed when my father arranged for Paul to join the Douglas-Hamiltons as their assistant. It was to discuss Paul's role that I first met the Douglas-Hamiltons just after I returned, over lunch at Nairobi's Intercontinental Hotel.

My father began that meeting by introducing Iain to me as "Douglas Iain-Hamilton," and I felt my cheeks burn. How *could* he have done that? Knowing Iain as I do today, he probably didn't hear or wasn't listening, his mind already racing ahead to embrace some fascinating idea that had captured his imagination. It was a typical meeting with the Douglas-Hamiltons, who always managed to be surrounded by an aura of drama and mystique. I was giddy with excitement, and to this day I feel the same in their presence. I was fascinated by Oria and wished I could be more like her, with her almond eyes, dark Italian beauty, and deep passion for life. Years would pass before I got to know Iain and Oria well, but eventually they and their two daughters, Saba and Dudu, would become like family to me.

Iain had just been appointed coordinator of the Elephant Specialist Group formed by the International Union for the Conservation of Nature (IUCN) and was undertaking the first continentwide census of African elephants. Paul was to assist in the continued monitoring of the Manyara elephants and to help Iain in his upcoming count of the elephants of the vast Selous Game Reserve in southern Tanzania.

There were signs that all was not well for the elephants, and Iain was attempting to assess their status across the continent. The price of ivory had recently begun to rise and with it came a wave of slaughter. Two years before, in 1973, a gathering of experts had estimated Kenya's elephant population at some 167,000. That seminar, organized by Peter Jarman, an official in the Game Department, had been called to look into the apparent abuses by the department involving elephant hunting licenses. It concluded that poaching was rampant in Kenya and that high-level corruption was responsible for the huge numbers of tusks that were leaving the country. Within two weeks of delivering his report, Jarman had been told that he was no longer needed and that his contract would not be renewed.

The situation had continued to deteriorate over the next two years, and it became clear that organized poaching was being carried out across much of the country. Soon after our return to Kenya, I recall standing with my father in the arrivals hall at Jomo Kenyatta International Airport when a short wiry man with piercing blue eyes walked over to greet us. My father introduced him as wildlife biologist Ian Parker, and then the two men moved away to speak quietly together. When my father returned to my side, he said, "That's a brave man. He's been instrumental in exposing the involvement and complicity of top government officials in the illegal poaching of elephants for the ivory trade. The newspaper which implicates the president's wife was impounded here at the airport and banned." I could tell from my father's tone that there was more to the situation than he was revealing. As I watched Parker walk away in his Indian-style sandals, his thin legs protruding from a baggy pair of khaki shorts, his hands thrust casually in his pockets, I was forced to face the fact that my intended study of elephants, the gentle giants, and an idyllic life in the bush was suddenly juxtaposed against chilling intrigue and danger, not caused by the elephants but by greedy men and women in the corridors of power.

Parker himself, as I later learned, was a controversial figure, a partner in Wildlife Services, a company that killed elephants for profit. It was argued that in some areas there were too many elephants and that they should be culled before damage to their habitat was irreversible. Yet I did not dwell on elephant poaching or culling, legal or illegal ivory sales or corrupt officials for long. Looking back now at the statistics and on the endeavors of friends like Iain and Oria, I find it hard to understand how I could have been so ignorant of the situation. I was young in spirit, idealistic, and carefree, and my father tended to protect me from the ills of the world. I put the thoughts of slaughtered elephants and corrupt officials to one side. I had unusual people to meet, places to visit, and things to accomplish. Little did I know then how the slaughter of the elephants would come back to haunt me.

Some time afterward I had the opportunity to visit Paul in Manyara, and I looked forward to spending some time with him alone. My parents had given us strict instructions that, when we were in their home, we were to set a good example for my younger brother and sister. A year earlier at the family dinner table, twelve-year-old Ginny had asked, "Where do you sleep when you visit Paul at Harvard?" There was an uncomfortable silence, and I stared hard at my vegetables until my father answered simply, "She sleeps with Paul. There is a hole in the ceiling, and God watches over them." Everyone laughed, for my father had never spent a day in church in his life. But what I now found in Manyara was hardly romantic. Paul had set up his two-man tent on the edge of the dry riverbed at Ndala, where he slept on a foam mattress without sheets and ate his food straight from the tin. He bought a few mangos and bananas in the town of *Mto-wa-Mbu,* River of Mosquitoes, and had an occasional beer at Mama Rosa's. And although his strong body was tanned a golden brown, his shorts were filthy and his tennis shoes rank. He watched elephants from Iain's old open-topped Land Rover and, with his cowboy approach, driving up to the elephants too fast and stopping too close to them, he was charged repeatedly. During my visit, a big female chased us flat-out across an open glade, and, while I screamed and the engine roared, she put her tusk through the back of the vehicle. Like Iain, Paul preferred to watch elephants on foot or from the vantage point of a tree, and he proudly showed me several hairs he had managed to pull from an elephant's tail. Yet, with all the time he spent being chased, climbing trees, or getting stuck in aardvark holes, Paul had in fact not identified more than a few elephants. It took too much time, he explained, to go through all of Iain's old photographs of the animals and match them to the elephants in the field. I was not impressed. I located Iain's box of elephant pictures and got to work.

With Paul observing elephants in Manyara, I set my hopes on joining Cynthia Moss's elephant study in Amboseli. Cynthia had gone to East Africa in 1967 as a tourist, met Iain Douglas-Hamilton in Tanzania, and resigned from her New York City job at *Newsweek* in order to assist him with his study of the Lake Manyara elephants. She, like so many before her, fell in love with Africa and decided to stay on. During the months she had spent in Manyara, Cynthia had become enthralled with elephants and was determined to start her own project.

In 1972, with assistance from the African Wildlife Foundation, Cynthia began photographing the Amboseli elephants and getting to know them individually. I was thrilled to learn that she also worked part time as the editor of *Wildlife News,* the magazine produced by the African Wildlife Foundation, because with my father the new director, there was a good chance that I would soon have the opportunity to meet her. If I was lucky, she might allow me to join her elephant study.

Several weeks later I was introduced to Cynthia at a party she gave to celebrate the publication of her first book, *Portraits in the Wild*. She lived then, with her smoke-colored cat, Moshi, in a series of tin huts and caravans on a ranch outside Nairobi, where she rode horses across the Athi-Kapiti plains. We drove there after dark, candles set in sand inside paper bags marking the way along the meandering dirt tracks, the pink eyes of spring hares dancing like fireflies in the night. Cynthia was a free-spirited woman and part of a large social crowd of wildlife biologists who were leading extraordinary lives in Africa. Sixteen years older than my nineteen, she was just taller than I with long blond hair; a product of the 1960s, she had worldly experience and a self-confidence I envied. She had a warm, infectious laugh that enveloped me with a sense of well-being. In October 1975 I went to stay with Cynthia in Amboseli and my life with elephants began.

CHAPTER ⇒ 3

The Elephant Camp

MY PARENTS DROVE ME to Cynthia's camp, set in a clearing in the shade of three enormous yellow-barked *Acacia xanthophloea* and bordered on three sides by a dense wall of dark green date palms and a thorny mass of young regenerating acacia. A cloud of dust followed us as we drove into the camp. Except for the occasional pieces of straw-colored stubble poking up through several inches of fine volcanic ash, the ground was bare. We pitched our tent under an old fever tree, in what thereafter became known as Poole Camp. Poole Camp was not a hundred meters from Cynthia's tent, but it was beyond the white rocks that demarcated the edge of the legal public campground, and I remember Cynthia's disapproving look when my father chose the spot. My parents stayed for the weekend and then went back to Nairobi, leaving me to embark on a new life. I moved my tent back across the white rocks into Cynthia's camp and pitched it near the dining room, east of the other tents.

Cynthia had established the elephant camp in September, only a few weeks before I arrived. Although researchers and tents came and went over the years, the basic structure of the camp remained much as I first saw it, until the floods that followed the 1988 short rains finally forced us to move to the higher ground of Poole Camp in 1990. The Elephant Camp, as it came to be known, was to become the home for members of the Amboseli Elephant Project, although, over the years, a number of scientists from other studies were invited to live there.

The camp was situated at the edge of the palm and acacia woodland known as *Oltukai Orok*, the place of the dark palms, and looked south toward Kilimanjaro over a glade we referred to as the South Clearing. A second glade, north of the camp, we imaginatively called the North Clearing. Elephants, buffaloes, impalas, and warthogs visited us every day, passing through the camp as they moved back and forth between the two clearings. The central feature of the Elephant Camp was the dining room tent, our "two-room canvas mansion," which was shaded by the three large *Acacia xanthophloea*. The dining room doubled as library and sitting room and was where we gathered to talk about the day's events over tea, at any time of day, or over a beer in the evenings. Each of us had our own tent, and these together with the kitchen, the shower, and the "long-drop," or outhouse, all radiated out in a circle around the dining room.

The camp kitchen stood in the shade of an enormous date palm. Although we would have several different kitchens over the years, they would remain similar in design to the one I found in camp when I first arrived: a small rectangular building with walls of sisal poles. The roof was thrown together from various odds and ends: bits of old plastic, roofing felt, old tents, and corrugated iron, or *mabati*. Lots of useful things that might have been stored in a kitchen closet had we had one "lived" instead on the roof, including bits of wire, rope, brooms, *mapanga* for cutting firewood, and slashers for cutting the grass. There were also numerous other things up there that looked as if they should be thrown away but that

Masaku Sila, the camp cook, insisted he had a use for. The roof was also a favorite place for snakes, and many deadly black mambas had dropped off from it, landing with a thud and thrashing violently past the washing-up table and into the palms.

Tacked to the front of the structure was a small sign that read EXERCISE CAUTION. The sign was mere decoration, but more than a few visitors assumed that the kitchen was home to some dangerous wild animal. With no windows except the small gaps in the sisal poles, the interior was cool and dark, and its earth floor felt luxurious against hot, bare feet after a long day in the bush. The contents of the kitchen were very basic but perfectly adequate: a small gas stove, a kerosene refrigerator, a couple of old tin trunks that held fruits and vegetables, flour and maize meal, or things that didn't belong anywhere else, and three cupboards made in nearby Loitokitok. The two smaller cupboards held cutlery, spices, jams, candles, dishtowels, lantern wicks, bread, spaghetti, and breakfast cereal. The third larger cupboard was backless and held the plates, glasses, teacups, pots and pans, and egg tray. Small mammals and reptiles found this cupboard, with its easy entry, a favorite place to visit, and pots that were used infrequently always contained rodent droppings. After a small black mamba was found curled up in the egg tray, we reached in there with some trepidation.

One of the most important items in the kitchen was the photo display, a collection of snapshots of memorable camp characters and events pinned onto a board and covered with a piece of old plastic. The pictures eventually included such events as Boniface, Masaku's assistant, with the eight-foot dead black mamba draped around his neck, Masaku with the "python that came back from the dead twice," Cynthia cleaning up the morning after Tuskless, the elephant, knocked the kitchen down, and me in traditional Maasai dress.

We had to make sure the kitchen door was kept tightly closed, because the vervet monkeys were always waiting for a chance to sneak inside to steal a piece of ripe fruit. In the early days, this required lifting the door and latching it shut with a rusty old padlock. Years later, when Wambua, the "engineer," joined us, he devised all sorts of clever gadgets to open and close the door "automatically." When a button was pushed the door popped open, after which you simply flung it to the side where it was caught and held open by a special latch. To close it you hit a piece of wood that lifted the latch and released the door, and it would slam shut. Even with these systems the vervets came to recognize Sundays as the camp staff's day off, and they invariably spent the day trying to get into the kitchen, defecating in our cars, and generally harassing us.

Outside the kitchen door was the washing-up table, and around the corner, next to the door of Masaku's tent, was a large pile of wine bottles, which were kept solely for the purpose of keeping the likes of Teddy, Taabu, and Tuskless, the kitchen-raiding elephants, at a safe distance. When they came too close, Masaku tossed one at them. There were very few camp rules, but one was very important: The camp must never be left unguarded for more than half an hour, for the elephants, too, monitored our comings and goings. They took advantage of our absences to raid the kitchen for the delicacies they loved, among them pineapples, bananas, tomatoes, cabbages, garlic, boxes of spaghetti, bags of maize meal.

To the left of the kitchen was the camp shower, which consisted of a twenty-liter aluminum *debe,* or tin can, hung from a horizontal post with a shower attachment welded to its bottom. To turn the water on we pulled down on one end of a seesaw handle; to turn it off, we pulled down on the other. The monkeys quickly learned to turn the shower on, and eventually we were forced to replace the handle with a tap. We could not afford to waste water; it had to be hauled from the *Ol Tukai* borehole several kilometers away twice a week in two hundred–liter drums. It was a chore we all loathed.

Water for the shower was heated over a fire in another two hundred–liter drum. The original method for getting water from the drum into the *debe* involved using a small foot pump and was designed by a fellow researcher who thought the design ingenious,

as did I, for a time. When the drum was full and the water hot, this technique was relatively efficient, but it was somewhat less so when the drum was less than half full, its usual state. Eventually we abandoned the foot pump and replaced it with a simple pulley system, which was much more satisfactory: The *debe* was lowered, water of the correct temperature was poured in, and the *debe* was raised back to its original position.

Surrounding the shower was a three-sided sisal pole fence for some semblance of privacy. The number of poles decreased with time, and those that remained were neither vertical nor parallel. With the evening sun illuminating our bodies, we felt very exposed. From inside the shower we could watch Masaku doing camp chores on one side and elephants in the palms on the other. Masaku made a practice of whistling when he had business near the shower, to alert us of his presence, just in case we hadn't noticed.

The *choo,* or long-drop, was thirty meters from the dining room tent, diagonally across from the kitchen, and deep in a grove of palm trees. It always seemed to me that the *choo* should have been situated with a view of the clearing, so that we could look out at the mountain or watch the elephants go by. Being in the palms, the *choo* was dark and seemed an ideal place for snakes. Checking for them was something we came to do as a matter of course. Because elephants, buffaloes, and lions frequently came through camp at night and their trail passed between the sleeping tents and the *choo,* it was not a place we ventured after dark unless absolutely necessary.

Each of us had his or her own tent, and my first was east of the dining room. It was large, ten by ten feet, and I slept on the ground in the very middle of it on a thin mattress. During those first few months I had very little else in the tent except two wooden orange crates in which I kept my field notes and my clothes.

CHAPTER ⇒ 4

The Five Males

ONE BULL RAISED HIS HEAD, ELEVATED HIS TRUNK, AND MOVED TO
FACE US. HIS GARGANTUAN EARS BEGAN TO SPREAD AS IF TO CAPTURE
EVEN THE SOUND OF OUR HEARTBEATS.
—Beryl Markham, *West with the Night,* 1942

OCTOBER 1975 WAS MIDWAY through the long drought in which
ten thousand of Tsavo National Park's elephants had already perished from malnutrition and starvation. In Amboseli, not 150 kilometers away, the elephants were more fortunate. In small groups,
the matriarchs led their families slowly back and forth between the
quiet shade of the *Acacia tortilis* woodlands and the cool of the
swamps. Though their thick, wrinkled skin hung from their gaunt
bodies in deep, leathery folds, most of the adults would survive as
they had the long, coarse *Cyperus papyrus* and *Cyperus immensus*
from the deep swamps to depend on. But the females' empty
breasts drooped withered from their chests and, without their
mother's milk and unable to feed on the thick swamp vegetation,
many of the babies and young calves would perish. It was the
fourth year of poor rainfall. The females' fat reserves were de-
pleted, and their estrous cycles and ovulation had all but ceased.
During the following year, the culmination of the drought, only
twenty-nine babies were born, and only fourteen of them sur-

vived; only one calf was born during the following two years. But I was new to the subtleties of elephant life. There were so many things to see and to learn that much of what I witnessed in those first few months in Amboseli passed by me almost unnoticed. With time, the only reminders of the difficulties the elephants faced are the photographs showing protruding pelvic bones, the long list of mortalities, and the striking dip in the age-structure curve, a missing cohort, that we will continue to see for the next 50 years, or until all of this age set die.

Cynthia had been working in Amboseli on a part-time basis since 1972, visiting the elephants when she could and steadily building up a photographic recognition file. In 1975 she moved to the park to begin her full-time study of the elephants, with a particular interest in the females: their social relationships, leadership, and demography. Female elephants live in family groups, and her first task was to get to know the members of each family. By the time I joined Cynthia she had fully identified and named many members of the central families, while others she was still in the process of getting to know.

In those early days I often went out to watch elephants with Cynthia, and it was from her that I learned how to recognize individual animals. Getting to know an elephant family requires considerable time and patience, and I watched as Cynthia carefully counted the number of individuals in a group, drew sketches of elephant ears, and attempted to take identification photographs of each adult female. A perfect elephant ID card contains three mug shots, one from the left side, one from the right, and one head on. Cynthia maneuvered her old Land Rover round and round the elephants, carefully positioning herself so that the sun shone directly on their ears, illuminating the warts, holes, tears, notches, and venation pattern that make each elephant unique. She then wrote down the film and frame number of each photograph next to the description of the elephant in her notes.

Initially all the elephants looked alike to me: large and gray with big ears. By listening to Cynthia, watching her careful note-taking, and studying the elephants' ears through my binoculars, I began to see how different each ear was. There were small ears, big ears, smooth ears, ragged ears, round ears, flop ears, curtain ears, and crumpled ears; there were ears with any combination of flap cuts, scoop cuts, V cuts, wedges, nicks, notches, tears, and dingleberries. Tusks, too, were different: There were elephants with long tusks, short tusks, splayed tusks, asymmetrical tusks, convergent and crossed tusks; there were elephants with only one left or one right tusk; and there were elephants with no tusks at all. As I encountered each new elephant, I memorized the pattern of its left and right ears and the shape and length of its tusks. With time I began to see that, just like people, each elephant also had a different body shape and face, and that there were strong family resemblances between sisters and between mothers and daughters.

Amboseli elephant families, which ranged from two to thirty individuals, were often in the company of other groups. During the wet season the elephants frequently traveled in aggregations of over a hundred individuals, and as they milled about interacting with old friends, getting to know family members could be an almost impossible task. Cynthia was fortunate in her study; the long drought meant that there was not enough food available for the elephants to socialize, and as a result they moved about in small, discrete family units. The habitat was open, and the elephants of central Amboseli were relatively tolerant of vehicles. There couldn't have been a better time and place to become acquainted with a population of elephants.

Although the female-led families of central Amboseli were well on the way to being identified and named, the independent adult males were still relatively neglected. Cynthia told me that she would be grateful if I spent some time sorting out the males because, she said, laughing, "They're boring." That suited me fine; I didn't think that I would find them boring, and I was pleased to have my own task to work on, my own feeling of importance. I had been vaguely worried that my father had coerced Cynthia into agreeing to have me assist her when she might rather have been in Amboseli on her own. "Sorting out the males" would give me a job that she didn't want and one that might ultimately be useful to the project as a whole.

Having left their families as teenagers, adult males don't remain in any one group regularly. As a result, they are more difficult to get to know than are the females. You cannot expect to find a male elephant with the same individuals or in the same general area as he was the first time he was encountered. You cannot assume that the biggest male in a bull group is inevitably the "leader" and that he will be accompanied by a particular set of "followers." Bull groups are not that predictable; as a result, if the animals are ever to be distinguished, the ears, tusks, body shape, and size of each must simply be committed to memory.

My father had lent me his camera, binoculars, and the family VW bus, and with my new responsibility I spent the mornings out with groups of bulls, taking identification photographs and making my own detailed drawings of their ears. In a small notebook I drew the notches, tears, and holes on the left and right ears of each of the adult males that I encountered. My earliest efforts looked something like this:

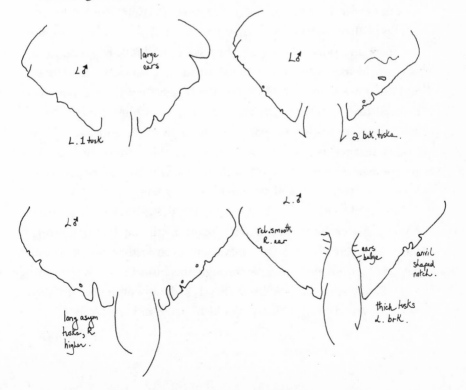

On one of my first days out on my own I drove over the Causeway that crosses *Enkongo Narok* swamp and left the VW bus to climb Observation Hill to look for some bulls to identify. From the summit I could see five large males coming from the *Ilmarishari* woodlands, striding across the open plains toward *Enkongo Narok*. By the time I had retraced my steps to the vehicle, they had disappeared into what was then thick *Salvadora* and *Suaeda* bush along the edge of the swamp. I drove off the road, through the foliage, and down to the swamp edge. This was my chance to work with my first elephants, and I perched quietly on the roof of the bus, carefully drawing all of the details of their ears.

As I sat absorbed in my work, one of the males began to move in my direction. I was still very new to the subtle facial expressions of elephants, and to me he looked distinctly aggressive. I quickly climbed back through the roof hatch, but it was clear that I had become the object of his attention, and he moved rapidly toward me with ears flapping and fierce eyes. I began to hope that he would lose interest, but he continued approaching until he was just three meters from the car. Suddenly, to my horror, the other four also noticed my intrusion and paced over, shaking their heads and swinging their trunks, until they had formed a semicircle around the front and sides of my car. They were big bulls, but with their heads held high and their ears extended they seemed even larger, and the more frightened I became, the bigger they grew. I suddenly realized that the roof hatch was still open, and one of them could reach in quite easily and grab me with a trunk or gore me with a tusk. I reached up slowly to close it, but the movement brought an immediate response from the elephants, who moved even closer. I left the hatch open and slid back into my seat.

My mind worked frantically, weighing the risks of fleeing versus holding my ground. I knew from many family safaris that starting the engine probably would annoy them even further, and it would have been difficult to escape through the bush in reverse, in any case. Panic-stricken, I realized that I had no option but to stay where I was. I judged that I would be safer and less likely to antagonize them if I crouched on the floor behind the front seat. As I hid the males came closer and closer until they were almost touching the car. I remember wondering what my parents would say when I returned home to tell them that the car had been flattened. Then it occurred to me that I might not return home at all. From where I hid, scarcely breathing, I could just see their huge heads looming over the car, two on either side, one in front. Curled up behind the seat, my legs were getting cramped. As I tried to change position the car squeaked, which after the long silence must have scared the elephants (whom I now realize must have been dozing). Suddenly I heard the sound of the huge animals crashing through the thick vegetation, and I crept up in time to see their tail ends disappearing into the bush. I trembled for two hours afterward.

I told Cynthia of my ordeal when I returned to camp. She laughed at me and said that they were probably just being friendly. To this day, I can still recall the elephants' expressions, but it was many months before I admitted that Cynthia was right and that they had come over just to say hello. Looking back through my drawings, I later discovered that one of the elephants was Aristotle, a docile older male who became one of my study males for many years.

During the first few months in Amboseli I continued to spend the mornings out getting to know the males and the afternoons in camp going through identification photographs. Cynthia had already cataloged sixty-eight males, but there were packets of additional photographs that needed to be sorted and compared with the original sixty-eight as well as with my own growing number of sketches. Pictures of left ears had to be matched with those of right ears, and the venation patterns of two left ears had to be examined with the aid of a magnifying glass. Did they belong to the same elephant or were they from two look-alikes?

Cynthia had pasted the black-and-white photographs of the first sixty-eight males onto pink manila pages in a looseleaf binder. Each male was given his own page and number, M1 (male number 1) through M68. I went through this Bull Book, as it was called, to try to match the photographs with my careful drawings to determine who was still alive and who was dead, and to ensure that each picture was actually of a different elephant. I discovered, for example, that M21 was the same as M51. In organizing the file, I realized that I had never seen most of the big males in the photographs and I wondered where they had gone. Only with time would I learn that they were dead, killed by hunters and poachers between 1972 and 1976. Many of these elephants had been enormous and clearly had been killed for their tusks. Of the original sixty-eight, only eight males over 30 years old remained, though the life span of an elephant is some 65 years. The largest and oldest of these were M13 (Iain), M22 (Dionysius), M28 (Cyclops), and M41 (Aristotle), with only Iain being over forty years old. Today only six of the original group are living: M5 (Sheik Zayed), M7 (Masaku), M10 (Oloitipitip), M22 (Dionysius), M48 (Ed), and M51 (Alfred).

I continued cataloging the males and soon reached M150 (Beach Ball, so named because of his rotund body shape). By 1995 the Elephant Project had registered over 450 males, and, thanks to the efforts of a number of researchers, their births, deaths, and family history are stored in a computer database. As the number of photographed males increased, it became more and more difficult to sort through the many packets of pictures and determine who was who. Out in the field I might meet a male I thought I recognized, medium size with a flap cut in the top of his right ear, a large hole in the middle of his left ear, and long asymmetrical tusks, but have no idea what his ID number was. By the time I had searched through the Bull Book and the collection of loose photographs, my subject had long since disappeared. Clearly the Bull Book was no longer practical, and Cynthia devised a system of pasting pictures onto computer punch cards, which became known, naturally enough, as the Bull Cards. Together we defined each one of the 50 or so holes around the border of the card as corresponding to a particular physical characteristic. For example, hole number 5 was "notch in the upper right ear," while hole number 33 was "asymmetrical tusks, left higher." Two photographs of each male, one showing the left ear and the other showing the right, were pasted on the front of the card, and one photograph of the male head with his ears spread was pasted on the back, and then the holes that described his particular characteristics were punched out. By pushing a long needle through the hole for "flap cut in the middle of the right ear," for example, and shaking the pile of cards, the few males who had a flap cut in the middle of their right ear would fall from the deck, making field identification more efficient. We used this technique until we could recognize each male without the aid of photographs. Since 1978 the Bull Cards have lived in a very solid wooden World War II field radio box, known as the Bull Box.

CHAPTER ⇒ 5

Naming an Elephant

CYNTHIA TOOK ELEPHANT NAMING very seriously and had developed a set of rules for her system. Since there were more elephant families in Amboseli than letters in the alphabet, Cynthia gave each one a two-letter code. Thus, there was an AA family and an AB family, a ZA and a ZB family, and so on. Each elephant in a particular family was given a name starting with the first letter of the code, so that, for example, every individual in both A families was given a name starting with the letter A. Cynthia tried to follow an ethnic theme when naming each elephant family. The WBs were given Kikuyu names, such as Wairimu, Wamaitha, Wangoi, Wambui, and Wanjiko, while the MBs were given Irish names, such as Maggie, Molly, Megan, and Moira. The RAs had Mediterranean names, such as Remedios, Raphaela, and Renata, and the UAs had the Scandinavian names of Ulla, Una, and Ulrica. With names like Gloria, Gladys, Geraldine, Grace, and Golda, I always imagined that the GBs were from New York City.

Cynthia also gave each female a three-letter computer code name: thus, BIG, stood for the BB's matriarch, Big Tuskless, BET for her sister Bette, BEL for Belinda, BON for Bonnie, BAR for Barbara, and so on. Initially Cynthia used the first three letters of the name for the code, but as the number of elephants increased, it became more and more challenging to make a unique code for each female. In her search for new names, Cynthia accumulated dozens of baby-name books, and she resorted to imaginative names like Qola, Qalypso and Qumquat. Adult females and juveniles over four years old were all named. (Calves under four years old are subject to high mortality and were therefore simply referred to by citing their mother and the year of their birth: For example, Ulla's calf born in 1976 was referred to as ULL 76 until she was four, whereupon she was named Ute.)

When I arrived in Amboseli, most of the adult females had been assigned to families but only the matriarchs and other prominent females of the central part of Amboseli had been named. I went out with Cynthia one day in 1976, south of Observation Hill, in through the regenerating acacias to the edge of the swamp, where we met some elephants that she recognized as the JAs. They were led by a beautiful female with long, curved, asymmetrical tusks named Jezebel. The many other group members were still referred to only by the film and frame numbers of their identification photographs, and Cynthia announced that she would name one of these females after me. This was a great honor, and I felt a surge of pride that Cynthia thought I was worthy of bestowing my name on an elephant. She studied her photographs, found what to her seemed an appropriate individual, and then searched through the group for her. Suddenly she pointed to a female standing close to Jezebel and said in a satisfied voice, "There, we'll call that female Joyce," and it was done. Joyce was a large, handsome female with thick, symmetrical tusks and a scar on her otherwise nondescript ear. I felt slightly disappointed by her choice, as I had wanted Joyce to be as beautiful as Jezebel, and she was not.

Over the years, as my ability to see the differences and similarities between elephants was sharpened, I began to see qualities of Jezebel in Joyce. They were, in behavior and appearance, undoubtedly sisters. Their resemblance could be discerned in the subtle curve of their backs, the way the lower portion of their ears curved in toward their faces, the angle at which they carried their heads, and the slightly downward curve of their mouths. Joyce was younger and would probably outlive Jezebel to become matriarch, carrying on an elephant legacy.

Although my interests lay primarily with the males, I tried over the next few years to become familiar with the many families of Amboseli. Cynthia introduced me to some of the other matriarchs and, with help from the photographic records, I began to memorize the ears of their kin and to learn where to expect to find particular families, and how each family associated and interacted with every other family.

Female elephant society is complex, consisting of multitiered re-

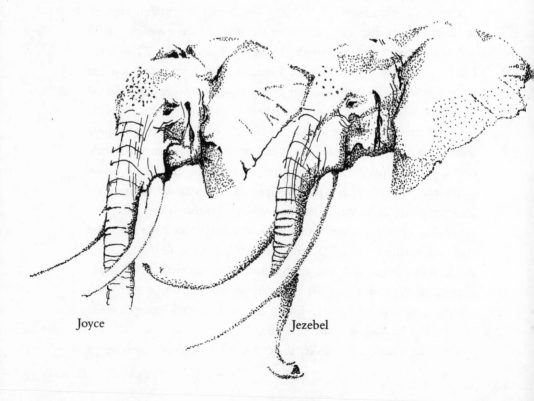

Joyce   Jezebel

lationships extending from the mother-offspring bond out through family units, bond groups, and clans. The basic social unit is the family, which is composed of one to several related adult females and their immature offspring, and may range in size from one to perhaps thirty individuals. For example, in 1976 Joyce belonged to a family of fifteen elephants whose family tree looked like this:

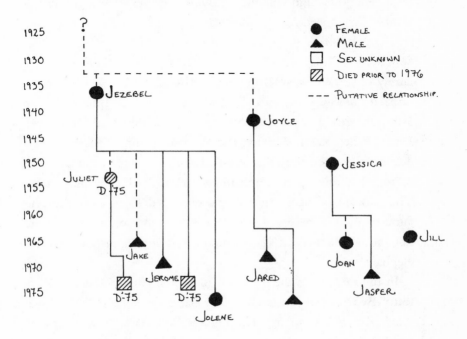

In his study of the Manyara elephants, Iain Douglas-Hamilton had discovered that elephant families typically have special relationships with one or more other families. Iain assumed that these families were related, and he referred to them collectively as a kin group. Cynthia, too, found that some families had a very close bond with one or more other families and that these elephants might spend as much as 80 percent of their time together. Although Cynthia believed that these special friendships between elephant families were usually based on kinship, she had observed at least one instance when she knew they were not and, therefore, preferred to call them bond groups. Cynthia learned that, like family units, bond groups show a high frequency of association over time, act in a coordinated manner, exhibit affiliative behavior (such as reassuring one another's babies, leaning against one another while resting), and form coalitions against perceived competition or danger. She also discovered that when family or bond group members meet, an extraordinary greeting ceremony occurs, during which the elephants may spin around urinating and defecating and, with their heads and ears high, fill the air with a deafening cacophony of rumbles, trumpets, roars, and screams.

The family that Joyce belonged to was bonded with two other families, the YAs, a family of three led by Yolanda, and the SBs, a family of six led by Sona. Jezebel's, Yolanda's, and Sona's families utilized *Enkongo Narok* swamp and land west beyond the *Ilmarishari* Hills, an area they shared with six other families led by Filippa, Freda, Isabel, Karen, Harriet, and Wartear. Cynthia referred to elephants that used the same dry-season home range as a clan, and so together these particular families made up the *Ilmarishari* clan. The concept of clans was a useful way to describe how different families utilized space and associated with one another, but we did not know whether the clan structure had any significance for the elephants.

In the mid-1970s the elephants of central Amboseli consisted of thirty-two families divided into four clans: *Ilmarishari, Oltukai Orok, Olodo Are,* and Southern. The *Oltukai Orok* clan consisted of nine families and was centered around the *Acacia xanthophloea* and palm woodlands of *Oltukai Orok* and extended to the eastern side of *Enkongo Narok* swamp and the northern end of *Longinye* swamp. The *Olodo Are* clan utilized the area from *Olkelunyiet* through *Olodo Are* to central *Longinye* swamp and was composed of five families. The Southern clan also included five families, it also utilized *Longinye,* but spent the nights in the *Acacia tortilis* woodlands to the south of the swamp toward the mountain.

The early and mid-1970s was a period of immigration, as the elephants sought refuge in the safety of the reserve from the hunting and commercial ivory poaching that was taking place outside. But there were other pressures on the elephants, too. In 1974 Amboseli had been established as a national park, and the Maasai, who had traditionally used the area's swamps in the dry season for their livestock, were unhappy with the turn of events. As a form of political protest warriors speared increasing numbers of elephants and rhinos. At the same time, the lifestyle of many Maasai was beginning to shift from seminomadic pastoralism to a more settled way of life, centered on the cultivation of maize and onions around the swamps that lay outside the new park boundaries. In their new role as farmers they were understandably less tolerant of elephants who trampled and devoured their crops.

Elephants were thus forced from their traditional ranges, and I frequently encountered "mystery females" and their families beyond *Olodo Are,* toward the large yellow fever trees of *Olkelunyiet* in the east or in long elephant grass in the west near *Lenkolong* and *Kitirua.* These terrified elephants usually ran away from the sound of the engine, their heads high as they looked back over their shoulders, the whites of their eyes showing, their tails curled high over their backs. Once in the tall grass, I couldn't see where I was going nor where I had come from, and sometimes these wild, frightened elephants caught me unaware, a whole family charging out of nowhere, *en masse,* trumpeting loudly, their great ears folded and flapping, their heads held high, looking down at me over long, sharp tusks. It was terrifying: the roars and rumbles coming from all sides, deafening even over the scream of the car engine, stuck on a tussock, as I searched for a way out.

It was there in the elephant grass that I first saw some of the new wilder groups that came to be known originally as the Peripheral Elephants and then later as the Western clan. I told Cynthia that through the dust raised by the charging elephants I had seen an unknown female with long, straight tusks and a left ear that flopped forward. She later became known as Bronwen, a member of the BC family, which, with the JBs and HBs, made up a bond group of three families led by the matriarchs Broken Two Tusks, Justine, and the majestic Horatia. They were part of a large clan of elephants that probably started moving in from Tanzania in the mid-1970s when ivory poaching in the Longido Game Controlled Area, south of the border, was rife.

Over time we got to know these wilder groups. Having learned that Amboseli was safe and the researchers harmless, they began to relax. Cynthia eventually identified thirteen families in this clan, led by matriarchs Horatia, Beda, Justine, Low Ears, Zipporah, Zorana, Athena, Galatea, Maggie, Nicole, Petra, Quilla, and Wangoi.

To the southwest of Observation Hill, along the edge of *Enkongo Narok* swamp, I met another family of wild elephants, who burst through the thick tangle of regenerating acacias at me. As they charged and then ran away I noticed a large female whose anal flap had swollen into a huge lump that dangled under her tail. The "flap-sac" female, as I called her, became known as Leticia, the matriarch of the LB family. Originally Cynthia grouped this female and her family with the other Peripheral Elephants, but eventually she decided that this elephant had in fact moved into central Amboseli from the east with four other families, led by Kora, Omega, Gloria, and Isis. Cynthia believed that these five families made up a clan that had once centered around the two swamps east of Amboseli, *Namolog* and *Kimana,* and that they had been forced to change their dry-season home range and move to the safety of Amboseli when the Maasai began to grow irrigated crops.

By the late 1970s all of the families in the Amboseli population had been charted, and by the early 1980s each of the adult males and females had been given individual identifying names and numbers. The population at that time numbered 615 elephants, consisting of 164 adult males and 451 adult females and calves, divided into six clans and fifty family units.

CHAPTER ⇒ 6
The Green Penis Syndrome

ONE AFTERNOON IN FEBRUARY 1976 a group of females came through the North Clearing and into the camp, followed by an enormous male whom I had never seen before. His symmetrical tusks were thick and beautifully curved, and he carried his huge head carefully, as if its great weight might throw him off balance. He walked through camp with his head and ears high, towering over the other males, the females, and the calves. I had never seen such a large elephant before, nor do I think I have ever seen one as immense since. As he passed the dining room tent I noticed that urine dripped continuously from his penis. I took a closer look through my binoculars and realized that there was something terribly wrong with him. The constant dribbling of urine had apparently caused the sheath of his penis to turn a greenish color from what seemed to be a nasty fungal growth. Studying the male again, I estimated that he must have been at least fifty years old. Perhaps old elephants become incontinent, I thought to myself at the time.

I saw the huge male again on the afternoon of February 22 feeding in the young regenerating acacia on the southern end of *Lon-ginye* swamp. I now noticed that the sides of his face were marked by a dark stain of secretion oozing from swollen temporal glands. The temporal glands, located just behind the eyes, were still an enigma to African elephant biologists, and I wondered what was causing them to secrete so profusely. He was still dribbling urine, and his penis was still green.

Ten days later Cynthia and I were out watching elephants together when we saw him for a third time, again with a group of females. Looking back through her notes, Cynthia noted that she, too, had seen him before, once on the twelfth of February with two other males and again on the twenty-third with a group of females. On both occasions he had been dribbling urine. We saw him twice again in the following week. In honor of his magnificence, I named him Zeus and gave him an identification number, M103. But we never saw Zeus again: The last in a line of huge Amboseli males, he, too, probably succumbed to the ivory poachers.

Several weeks later I came upon another big, beautiful bull whom I had never encountered, standing with a group of females in a small patch of regenerating acacia near a favored bull area I had named Place of the Bulls. He, too, had a green penis and secreting temporal glands. He, too, was walking tall and sedately, closely following a group of females. He was in his late thirties or early forties and had long, curved, asymmetrical tusks. There was nothing remotely old or senile about this male. Cynthia and I saw him several times in and around *Oltukai Orok,* in the company of different groups of females, and once showing particular interest in an estrous female. I gave him a number, M117, and named him Green Penis, for his terrible affliction. After a few more sightings, just as suddenly as he had appeared, Green Penis disappeared again.

Long discussions developed in camp about the elephants' malady. Other researchers and visitors suggested that the green penis was a symptom of a disease, perhaps a form of elephant venereal disease. This seemed plausible; certainly both Zeus and Green Penis had been with females and appeared to be sexually active. For a time, the affliction became known as Green Penis Disease, or GPD, and although we joked about it, I got the impression that the possibility that GPD might be sexually transmitted made Cynthia slightly concerned about *her* females and me, in turn, defensive about *my* males.

In September 1976 I returned to the United States to continue my education at Smith College. Cynthia wrote to me with Amboseli news, telling me that during the first half of 1977, she had recorded a series of additional males with Green Penis Disease: first Dionysius, then Aristotle, followed by Agamemnon, and David. Each of these males had been in association with females.

When I returned to Amboseli for several weeks during the summer of 1977, Cynthia reported that some of the males with Green Penis Disease were extremely aggressive and warned me to be particularly wary of a male with a deep V cut in the lower part of his right ear. She had been out watching a group of elephants south of Observation Hill, when this huge male came and towered over her car in a very threatening manner. According to her notes, this new male with the V cut had temporal gland secretion. She moved off to another part of the group but, to her dismay, he followed her and went for her again, this time at full charge. It was unusual behavior for the usually placid elephants of central Amboseli, and she was more than slightly unnerved. As she drove away, she remembered back to her days in Manyara and decided to follow Iain Douglas-Hamilton's advice and to teach this bull a lesson: The next time he charged, she would charge him back. So, when for the third time he began to move toward her, she drove her Land Rover straight at him. To her horror, instead of turning and running away, he came for her at a full charge. They were on a collision course, and it was clear that the bull was not going to give way. At the last second Cynthia veered off across the open plains, her foot flooring the accelerator.

Deciding it best to avoid the nasty bull completely, Cynthia moved to watch another group of females who had gone off in the direction of the swamp. She drove across the plain and found a way through the line of young acacias down to the edge of the swamp. In those days the vegetation was a thick, tangled mass of thorns and branches, and there were only a few places wide enough to allow a vehicle through and onto the strip of grass along the swamp edge. Engrossed in her note-taking, she forgot about the big bull and sat quietly watching the elephants for close to an hour when she felt a shiver run down her spine.

Instinctively she looked in the rearview mirror; a pair of thick tusks and a huge gray mass filled the view. She knew immediately that this was the same animal and that he had gone out of his way to follow her down to the swamp. Without a moment's hesitation she started the engine, and, with the male charging after her, she found her way through the thick trees and out to the safety of the open plains on the other side. Later, when she had calmed down, she went back to look at the path she had taken: There she found his huge footprints in between the tracks of her tires and a squiggly line made by his trunk as it moved back and forth following her scent down to the swamp edge. The male with the ⦂ cut had clearly established his dominance, something he never again let any of us forget. Over the years his behavior became so legendary that he earned the name Bad Bull and was given the reference number M126.

During that same summer I began to notice that my favorite males were behaving in what seemed a very strange manner. Although Cynthia had written to me that Dionysius, Aristotle, Agamemnon, and David had been in the company of females, when I returned to Amboseli I discovered the first three completely uninterested in sex, choosing instead to spend their time with other males resting and feeding around *Oltukai Orok*. And I eventually found David, whom Cynthia said had completely disappeared, in *Olkelunyiet* on the eastern boundary of the park, also with a group of males, but ones I had never seen before. That summer Bad Bull and Cyclops instead were chasing after females and threatening other males.

Could it be, I began to wonder, that the Amboseli males had

individual sexual cycles? In Cynthia's chapter on elephants in *Portraits in the Wild,* which summarized what was known about African elephants at the time, there was no mention of it. In those days I was rather intimidated by Cynthia, who knew so much more than I did about elephants and about life, and I pondered my theory for a long time before I ventured to suggest it to her. She listened, but told me that many people had studied elephants, and none of them had found any evidence to suggest that males had sexual cycles. I was disappointed, but I still believed that there was a distinct possibility that they did.

I returned to Smith College for the fall of my junior year, and Cynthia later wrote to tell me of two more males with the green penis affliction, first Hulk, then Green Penis himself. As with the others before them, each of these males was seen in the company of females and behaving very aggressively toward other males. By this time Cynthia and I had decided that GPD was not a disease but a behavioral syndrome, so we renamed it GPS.

I returned to Kenya for Christmas and spent January 1978 in

Amboseli. There I found Dionysius, a male I had seen earlier only with other males, accompanying females and with Green Penis Syndrome. Just before I departed for Smith, I attended a lunch party at the National Museums of Kenya, where a friend of Cynthia's, Harvey Croze, handed me a paper that he thought might interest me. It was entitled "Plasma Testosterone Levels in Relation to Musth and Sexual Activity in the Male Asiatic Elephant." I flipped through its pages until suddenly a photograph caught my eye: It pictured a male Asian elephant marked with two large arrows. The caption read: "Asiatic elephant bull in musth. The male is fastened to trees by chains around his front and hind legs, and is showing some discharge from the temporal glands behind the eye (arrowed) and a dribble of urine from the penis (arrowed)." I knew immediately that we had made a very exciting discovery in Amboseli. Our males with Green Penis Syndrome were neither incontinent, nor were they suffering from elephant VD, they were in musth— and we were the first to document it in African elephants.

I had heard about musth and knew that among Asian elephants it was a period of extreme aggression that occurred once a year in adult males, lasting a few weeks or months. It was associated with a dark secretion that oozed from the temporal glands; only male Asian elephants exhibited this secretion, and it occurred only during the period of musth. For this reason the temporal glands themselves were often referred to as musth glands. Musth was assumed to be a period of intense sexual and aggressive activity, or rut.

While musth was well documented in domesticated Asian elephants, strangely it was thought not to occur in African elephants. All the pioneering researchers of African elephant behavior and reproductive biology (including Roger Short, Erwin Buss, Silvia Sykes, Ian Parker, Richard Laws, Harvey Croze, Keith Eltringham, and Iain Douglas-Hamilton) had looked unsuccessfully for musth in African elephants and concluded that it did not occur. They based their conclusions on the fact that while temporal gland secretions were observed frequently in African elephants, unlike in their Asian cousins they occurred year round and in animals of all ages and of both sexes. The secretion appeared, therefore, to be totally unrelated to sexual and aggressive activity. What they had not realised was that African elephants exhibit more than one type of temporal gland secretion, and that the secretion by males associated with urine dribbling was, in fact, musth. (The African elephant specialists had also become lazy with their terminology and referred, nonetheless, to temporal gland secretions in males or females as "musth" and any elephant with secretion as "having musth" or being "in musth" or "on musth," which had complicated the matter still further.)

For some reason most African elephant biologists had overlooked the dribbling of urine. While Iain Douglas-Hamilton had mentioned it in *Among the Elephants*, he did not make the connection between urine dribbling and the period of musth. Looking back through her notes, Cynthia found a record from 1974 when she, too, had commented that Cyclops had been dribbling urine.

Our discovery that African elephants did come into musth was very exciting, and I made plans to spend the summer of 1978 in Amboseli collecting data on musth for my honors thesis at Smith. It was the first real contribution I had made to the body of scientific knowledge on elephants, and I was terribly proud and eager to continue.

CHAPTER ⇒7

The Lasting Sorrow

YOU WOKE UP IN THE MORNING AND THOUGHT:
HERE I AM WHERE I OUGHT TO BE.
—Isak Dinesen, *Out of Africa,* 1938

I HAD JUST COMPLETED my third-year exams in the middle of May; in five days I would be returning to Kenya to see my family and to spend the summer watching elephants. I longed for Africa, the vistas, the smells, and the music. In Africa I woke up in the morning with a sense of purpose. I had my tent, my elephants, and the sense of freedom that comes with living in wide open spaces.

I had just turned twenty-two and the eager anticipation of my youth was heightened by the spring flowers, the warm sunshine, and the fact that I was with Paul at Harvard. We had been to several parties, one on a rooftop with old friends that my brother had come down from his New Hampshire school to join. Bobby had departed only that morning.

I was waiting alone in Paul's rooms for him to return from a lecture when the telephone rang. I picked up the receiver. There was an odd silence on the other end, some strange sounds, and then a trembling voice said, "There has been an accident, and your mother is alone." It was my grandfather, my father's father. He was unable to say more, and as I grasped the terrible truth of what he was trying to tell me, the room began to spin. I ran from the building seeing and hearing nothing. Outside the sunshine, the crowds, and the traffic were a blur. I was aware only of my feet slapping the pavement and of my mind screaming to shut out the world. I was searching for one face in the crowd: Paul's.

I boarded a plane for Kenya that night with my brother.

My father, we learned, had left the house after dinner, telling my mother that he had to make a telephone call to Malawi from the office. We could only deduce that afterward he had chosen to visit a friend on the other side of town. It was late, he was in a hurry, and he was driving too fast. The highway to Embakasi was under construction, and typical of third-world construction projects, there were no warning lights, no reflectors, no signs telling travelers that the road was about to diverge and become a divided highway. Only a few oil drums strewn about on the other side of the railway bridge indicated the detour ahead. Perhaps my father's mind was else-where. Perhaps he was blinded by the bright lights of an oncoming truck. Whatever the reasons, he failed to take the detour. His car was found underneath the truck. He had died instantly. He was only forty-five.

It was, it still is, incomprehensible to me, that a man so full of life could have been snatched away so suddenly, so irrevocably. All my dreaming, my desperate hoping couldn't make it less final. He was gone.

On that day part of my own spirit died, too. My father had meant the world to me. I was very much his daughter: Dark in complexion, with deep-set eyes and chiseled features, we looked very much alike. He was warm and friendly, and though he ap-peared totally at ease with people, he preferred the company of his family and was perhaps happiest walking quietly in the forest look-ing for birds. He cared deeply about making the world a better place, and he did everything with great energy and determination; there was so much he hoped to achieve in his lifetime that for him (and so, too, for us) wasting time was a sin. Perhaps most important for me, he was a man of principle who stood up for what he believed in.

While I am like my father in my outlook on life, I am more like my mother in temperament. An emotional child, I was quick to laughter, but also quick to anger and easily wounded and saddened. With my father's spirit of adventure and sense of principle and my mother's emotions, I frequently became deeply upset by the injustices of life. I think that my brother and sister would agree that my father protected me and tolerated emotional outbursts in me more than he did in my siblings, always taking me aside and talking everything through. He was always there when I needed him, and I adored and depended on him. By my early twenties we had reached the stage when father and daughter come to know each other as friends, and after his death I spent many years looking for someone to fill the terrible emptiness his passing left in my life.

As my father would have wished, there was no funeral, no memorial service. My mother, brother, sister, and I packed the car and, with his ashes, drove south through the Maasai Mara Game Reserve and across the Sand River into Tanzania. We drove through the woodlands and across the vast plains of the Serengeti National Park to his favorite place on earth, the Moru *kopjes*. Most would find it a lonely resting place, but for my father it would be paradise. We searched high among the rocky outcrops until we found one with a view in all directions. Together we tossed his ashes over the cliff and left a plaque on the rock on which I wrote the words:

> *He will hear the birds sing*
> *And the hyenas call*
> *He will hear the thundering hoofbeats of the wildebeest*
> *The sun's warmth will pour down on him*
> *And the wind and the rain*
> *Will make him a part of the earth he loves.*

Many years would pass before I would return to that *kopje* in the Serengeti, but my father was never far from my thoughts. Yet his death had shattered our family. It was as if my father, his ideals, his way of seeing the world, had been the inspiration and purpose that bound us together. From that day forward we changed forever, each of us trying to pick up the pieces of our lives.

In a sense my father had given Africa to me, and deep within me I felt a desire to continue living a life that he would have enjoyed. While my mother, my brother, and my sister returned to the United States shortly after my father's death, I made plans to stay on in Africa with the elephants.

PART⇒TWO

FIELD BIOLOGIST 1979–1984

CHAPTER ⇒ 8

Designing My Study

MY MEMORIES OF THE months following my father's death are mixed. Some events stand out sharply, others have been lost to a blur of grief. Although I had been torn between staying in Nairobi to help my mother pack up the house and trying to get on with my own life, in the end I decided to return to Amboseli, a choice I think both of my parents would have approved.

During that summer I saw six different males in musth: Iain, Green Penis, Bad Bull, Ajax, Richard, and Sleepy. I took careful field notes on each male and transferred them into a big blue analysis book. By carefully organizing the pages, I could quickly determine whether a particular male had been with females or with males and whether he had been in musth. As time passed and the number of entries grew, clear evidence of a sexual cycle for each male developed. I spent most of my stay observing Bad Bull, who was in musth for the three months I was in Amboseli. Watching the elephants and being chased by Bad Bull kept my mind occupied and were probably the best things I could have done to come to terms with my loss.

At the end of July 1978, my mother, brother, and sister left Kenya to return to the United States. I followed them a month later, stopping at the Subdepartment of Animal Behavior of the University of Cambridge to visit Professor Robert Hinde. By that time I had accumulated eight months of data that, when combined with Cynthia's observations, was beginning to show an intriguing picture of male elephant behavior. The more I learned, the more questions I had. I decided to continue my study of musth, and it seemed to me that I might as well earn a Ph.D. for my efforts. Robert had been the Ph.D. supervisor of two famous female primatologists: Jane Goodall, who studied the chimpanzees of Gombe Stream in Tanzania, and Dian Fossey, who studied the mountain gorillas of Rwanda. He also had supervised many of the interesting people I had come to know in Kenya. I was convinced that this was the man under whom I wanted to pursue my Ph.D.

I walked into Robert's office at the Subdepartment of Animal Behavior, my big blue analysis book under my arm, and I told him about the discovery of musth, showing him detailed examples of the behavior and symptoms of my males. A tall, lean man with a deep, compassionate voice and a shock of white hair, Robert listened to me with interest but explained that his research now involved human behavior and that he was no longer taking on students of animal behavior. He recommended instead that I speak to Keith Eltringham in the Zoology Department, who had studied elephants.

I was disappointed by Robert's remarks because I was interested in behavior and Keith, I knew, was an ecologist. In addition, I had read some of Keith's work and it was clear that he believed that musth did not occur in African elephants. While this was not surprising, as it was the general consensus at the time, it meant that I would have to convince him that I knew what I was talking about. Keith listened to me very attentively, but after I finished he explained to me in no uncertain terms that musth did not occur in African elephants. I left Cambridge hoping that Robert would change his mind and take on one more student of animal behavior.

I returned to Smith College in September to write my honor's thesis on musth. In many ways it was my most satisfying year at Smith because I was very excited about elephants and I enjoyed having my own project to focus on. Dr. Betty Horner, professor of zoology, took me under her wing and encouraged me to submit a student paper to the American Society of Mammalogists. To my surprise, I won an award for my paper and was invited to present it at the society's summer meeting in Oregon. I graduated from Smith in May of 1979 with high honors for my elephant work and a Smith College graduate fellowship. But best of all, Cambridge University and King's College wrote to say that I had been accepted as a Ph.D. candidate with Robert as my supervisor.

That summer I traveled to the mammalogist meetings where I presented my paper on musth. One of the earliest biologists to study African elephants was in the audience and asked a question. Although I can no longer recall what it was, I do remember that because of his conviction that musth did not occur in African elephants, he had missed the entire point of my talk. I felt quite exasperated. Although most field biologists had accepted my findings, it was many years before zoo biologists and veterinarians would be convinced of the occurrence of musth in African elephants and particularly its relationship to a male sexual cycle.

In October, I returned to Cambridge as Robert's last student of animal behavior. While I never knew what made him change his mind, I have benefited enormously from his decision. I would spend Michaelmas Term in Cambridge; just enough time to discuss my study with colleagues, plan my data collection, design my data sheets, and gather together some reference papers before heading to the field. During this period Cynthia and I also started to put together a paper on our discovery entitled "Musth in the African Elephant," which was published in the August 27, 1981, issue of the scientific journal *Nature*.

During my three-month stay in Cambridge, I had a small room in King's College, and every morning I cycled the five miles to the Subdepartment of Animal Behavior located in the village of Mad-

ingley. Many young men and women whom I had met in East Africa were now in Cambridge: Richard Wrangham and Phyllis Lee, who studied vervets in Amboseli; Jeanette Hanby and David Bygott, who studied lions in the Serengeti; Sandy Harcourt and Kelly Stewart, who were studying gorillas in Rwanda; John Scherlis, who was doing a study of the Manyara elephants; and Richard Barnes, who was writing up his study of the elephants of Ruaha National Park in southern Tanzania. I was to meet many more as we gathered at seminars, at parties, and in pubs to discuss our experiences and our discoveries.

At the Cambridge University library I sought as much material as I could find about musth in Asian elephants. The word "musth," I learned, comes from the Urdu *mast,* meaning "intoxicated." Musth in domesticated Asian elephants has been documented for centuries, and some of the early reports from the owners of working elephants were fascinating. Daily rationing of food and water could prevent the onset of musth, and if it was too late for prevention, remedies to minimize an elephant's rage included huge doses of Epsom salts or opium. In a book written in 1901 on the treatment of elephant diseases, G. H. Evans advised the following to calm a musth male's excitement: *"Four to six drachmas of opium or ganja (marijuana) given with boiled rice, plantains or jaggery; or three drachmas of camphor and two of opium twice a day for two days; or eight pounds each of wheat flour, onions and sugar and four pounds of ghee* [clarified butter], *mixed together and worked into orange-sized pills and administered one each night and morning until the whole is taken."* After this treatment, even the most troublesome animals would carry on their work as usual, according to Evans.

Early studies also speculated why musth occurred. Some writers claimed that they had observed a yellowing of the skin during musth and that this was an obvious indication of a liver disorder. Others argued that musth was a sexual malfunction, and that if a male were given access to a female, the symptoms would disappear. They cautioned, however, that if the female seemed to increase the animal's rage, she should be removed immediately. More recent scientific papers described musth as a form of rut and gave detailed description of its physical symptoms, the behavior of affected males, and the timing and frequency of musth periods.

At that time I knew of only eleven males in the Amboseli population who had come into musth: Bad Bull, David, Dionysius, Iain, Agamemnon, Cyclops, Hector, Ajax, Aristotle, Oloitipitip, and Green Penis (who had been renamed Harvey). I decided to concentrate on these particular individuals, but I also wanted to compare their behavior with that of some of the younger males who had not yet shown signs of musth and to watch how the behavior of these younger males changed as they matured. I chose the next nineteen oldest males for this purpose. These thirty males together became known as my "focal animals," that is, the elephants whom I would study intensively.

After discussing the objectives of my work with a number of people in the subdepartment, I decided that I would attempt to carry out at least one half-hour "focal follow" on each male each month during which I would collect detailed information on the male's physical state, his associations, and his behavior. I designed my data sheets in a way that would allow efficient information collection and analysis. At the top of each sheet I would record the date of my observation, time, location, the male's identification number, and whether he was in musth or not. If in musth, I would note the consistency and amount of his temporal gland secretion, how fast he was dribbling urine, how green his penis was, and what sort of general physical condition he was in. I would note the number of elephants in the group and list, by code name, all of the families present and the identification number of every adult male in the group.

Below this introductory information I drew in a series of rows and columns so that I could collect data on the proportion of time males spent engaged in different activities, their so-called activity budget, and take account of the rates of certain types of behaviors. I divided my data sheet into five-minute intervals and subdivided each of these intervals with narrow lines representing thirty seconds.

To measure the amount of time musth and nonmusth males spent engaged in different activities, I would note an individual male's activity every five minutes, on the minute. I defined eight different possible activities: feeding, moving while feeding, walking, interacting, resting, standing, drinking, and comfort behavior (for example, dusting, mud splashing, and so on). I would also note, every five minutes, the name of my focal male's nearest neighbor and estimate how close together they were in meters. Finally, I would make notes on the male's interactions with other elephants as they occurred. To save myself from having to write copiously, I defined a series of codes for individual males, for activities, for physical and sexual states, and for types of actions and responses. For reference, I drew a picture of Beach Ball to illustrate temporal gland secretion in amounts 1, 2, 3, and 4, and pictures of Iain to depict the degree of temporal gland swelling, numbered from 0 to 3 (see facing page).

Aggressive interactions between a focal male and his rivals or sexual interactions between a focal male and an estrous female can take place very quickly, and codes also helped me to keep pace

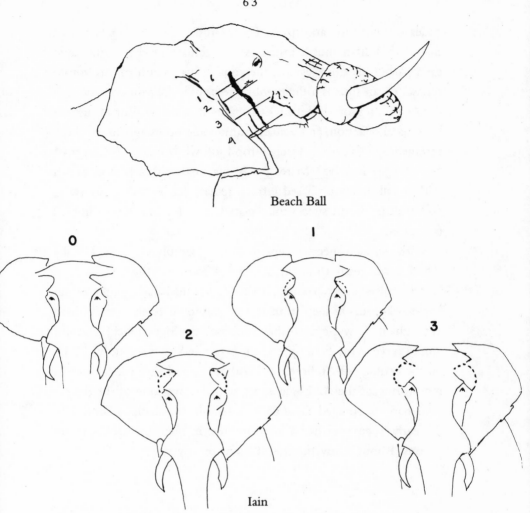

Beach Ball

0   1   2   3

Iain

with my focal male's activities. Thus, 126 EW MR EF T3 45 V3 (30) indicated that Bad Bull (126), in musth, ear waved (a threat), musth rumbled (a vocalization), and ear folded (another threat) as he walked toward Patrick (45), also in musth, who avoided Bad Bull by walking away when Bad Bull was still thirty meters away from him. I reserved a column titled "agonistic" for all aggressive interactions.

Under another column I would note all of the sexual interactions in which my focal male was involved. Again I defined a series of codes to save time and space. For example, 126 T1 Jez St meant that Bad Bull in musth tested estrous female Jezebel's vulva and then placed his trunk against the roof of his mouth on his vomeronasal organ (this is a flehmenlike response; the vomeronasal area is part of a sensory system which exists in many mammals distinct from nasal olfaction, and sexually stimulating odors are among those received by this organ); Jezebel stood still while being tested. 126✔ 1–2 CH Jez RA; 126 Mates Jez R meant that Bad Bull in musth with a full erection chased estrous female Jezebel who ran away; Bad Bull in musth successfully mated with Jezebel, who rumbled (vocalized).

With my data sheets complete and my sampling protocol established, I was ready to go to the field. I had my father's Zeiss 10 × 50 binoculars, a new camera, a new tent, and I had bought an old two-stroke Suzuki jeep from fellow Amboseli researcher Richard Wrangham. It was agreed that Amboseli ecologist David Western would serve as my field supervisor. David had carried out his Ph.D. on the ecology of Amboseli and had continued with the long-term monitoring of the Amboseli ecosystem. He was one of the people who had encouraged Cynthia to establish the elephant project in the early 1970s and had a keen interest in elephant habitat interactions. I bought my ticket and I was on my way.

CHAPTER ⇒ 9

Male Watching

IT WAS JANUARY 1980, eighteen months since I had left Kenya and twenty-one months since my father had died. Now I was returning to Africa on my own. Filled with excitement and anticipation, I boarded the plane. Eight hours later we landed in Nairobi. The doors of the aircraft opened and I breathed in the fresh air of an African dawn. I was home. But as my turn came to pass through immigration, my worst fears were realized: The officials would not let me into the country with a one-way ticket. Finally, after much protesting, floods of tears, and a ticket purchased out of the country to Ethiopia, I passed the barrier into Kenya and found Cynthia there to meet me.

The Amboseli that I returned to was a different place from the one I had left behind. The long drought had finally broken in late 1976, and several years of higher-than-average rainfall had followed. In early 1977 the females began cycling again, and twenty-one and a half months later, the first babies of a new cohort had been born: five in late 1978 and another fifty-seven in 1979. After almost two years of no births, suddenly 11 percent of the population was under two years old and the elephant social scene had changed greatly. With babies to take care of again, the females were more demonstrative and vocal. There was plenty of food for all, and families again joined to form bond groups, and bond groups joined to form clans. The elephants now moved in aggregations of several hundred individuals, rather than in the small family fragments that I had grown accustomed to seeing.

I left camp early each morning in my jeep searching for my focal males. I attempted to follow each one for at least a half-hour each month. For those males who came into musth my aim over the eighteen-month study was to collect approximately the same number of hours on each male during both his musth and nonmusth periods. I soon discovered that this was impossible: Some males were easy to find, while others disappeared for weeks, even months at a time. I could locate Iain, for example, on almost every day of the three months that he was in musth, but once he dropped out of musth he disappeared across the border into Tanzania. I also learned that the musth periods of some males lasted for months, while those of others lasted for only a few days. Musth is a *heightened* period of sexual and aggressive activity, thus males can be sexually active (and even mate) but not necessarily be in musth. Harmon, one of my younger focal males, lived out near *Kitirua* when he was sexually inactive, and I could find him there relatively easily, but though he was sexually active for several months, he was in musth only for a day or two.

Each male had a specific "bull" or "retirement" area where he fed and rested during his sexually inactive period. These bull areas tended to contain a higher biomass of woody vegetation than did the areas used by family groups. There were three different bull areas in Amboseli: *Oltukai Orok, Kitirua,* and *Olkelunyiet.* Each lay along the southern boundary of the park. As I accumulated data, I was able to determine the bull area for each of my focal males.

It was much more difficult, however, to assign the younger males to a bull area. As was true of other elephant populations that had been studied, Cynthia had found that young Amboseli males leave their natal families at around fourteen years of age. Newly independent males then follow one of several different courses to social maturity. Some young males abandon their own families only to join up with another family for several years; others go off to bull areas where they associate more loosely with bull groups; while still others remain in female areas moving from family to family. It is still not clear what influences a male in his choice of bull area. Some young males stay on in the general vicinity of their mother's clan area, while others move to the opposite side of the park from where they were brought up. It appears to take young males a decade or more to "decide" where they will settle, but once they have selected a bull area, most remain faithful to it for years, perhaps life.

I found that males who used the same bull area were often seen in close association with one another. Male elephants are in general much less social than female elephants, and it is difficult to say whether such groupings represent a special relationship between the elephants or whether they spent time together just because they happened to use the same area. My impression was that although the general pattern of association between males of a bull area was random, some of the older males established "friendships" with one another and that this relationship formed the basis of their close spatial association. Interestingly, some of the closest relationships were between males who looked very much alike, and I often wondered whether they had originally come from the same family or bond group.

Watching sexually inactive males in their bull areas was often tedious, and I spent many hours sitting on the roof of my Suzuki straining my eyes through binoculars every five minutes to see whether, for example, a lone Aristotle was still feeding, or just standing, in the swamp. Observing sexually active males, particularly those in musth, was much more exciting, and interactions between elephants often occurred faster than I could keep track of them. On one memorably chaotic morning I saw twelve matings involving both musth and sexually active nonmusth males in an hour; on another I was able to watch the complicated dynamics between three estrous females and five musth males in a single group. Even lone musth males were interesting subjects. They paced back and forth, rumbling and listening and dragging their trunks across their temporal glands; they marked trees with their temporal gland secretions and wallows with their urine; they tracked estrous females and other musth males; and they charged me not infrequently.

When I was not out watching or looking for male elephants, I worked in my tent, collating and analyzing my data. As this was before the days when biologists took laptop computers to the field, I had to buy another big blue analysis book in which I summarized the data from each focal follow. I allocated three pages for each male: one for data gathered when the subject was with other males, one for when he was alone, and a third for when he was seen in association with females.

In another smaller notebook I recorded every aggressive interaction that I had seen between any of my focal males. As these data accumulated I was able to organize them in three different matrices: interactions between nonmusth males, interactions between musth males, and interactions between a musth and a nonmusth male. On three large sheets of graph paper I listed each of the adult males along both the vertical and horizontal axes of each matrix. With the rows indicating the winners and the columns the losers, I entered one mark in the square for each day that I had seen two particular males interact. If the interaction had escalated to a fight, I made special note of that with a small f.

These charts eventually revealed that body size was a very important factor in determining dominance rank. Elephants are very unusual among mammals in that their growth in height continues long after puberty. Among females, growth levels off at around twenty-five years of age, ten to fifteen years after puberty, but among males, growth in height may continue until an individual is forty-five years old. Recent studies have shown that the delayed epiphyseal fusion of the long bones makes this growth pattern possible in male elephants. The epiphysis is part of the bone that ossifies separately and subsequently becomes fixed to the main part of the bone. The fact that male elephants break from a typical mammalian growth pattern indicates that through evolutionary time, large males have enjoyed a significant reproductive advantage over smaller males.

I used several different techniques to assess relative body size. Each time I observed two males standing together, I noted who appeared to be tallest at the shoulder and entered this information on a matrix. In another little notebook I collected a second measure of shoulder height: hind footprint length, measured from the back of the impression of toenails at the front of the foot to the end of the wrinkle pattern at the rear of the foot. Earlier work in Amboseli by David Western and others had shown that the hind footprint length of elephants is directly related to shoulder height. With these two pieces of information, I slowly built up a ranking of the relative sizes of my focal males. These data were to become very important as I tried to unravel the complex dominance relations between male elephants.

A third factor that I thought might influence dominance rank was physical condition. To gauge body condition, I gave a condition score of 1 to 5 at every sighting of my subjects, and I took photographs of each male, from the back, at the beginning, and at the end of his musth period. By measuring the ratio of a male's width to height on the photographs, I was able to show quite clearly that males lost weight during their musth periods. The condition scores showed that the longer a male was in musth, the more condition he lost.

As the number of focal hours I collected increased, distinct patterns of behavior began to emerge. One of my first observations was how dramatically the demeanor of the males changed once they came into musth. It was almost a Dr. Jekyll and Mr. Hyde transformation. Out-of-musth males were calm and amicable, and spent the majority of their time feeding or resting in their bull areas in

the company of other males; as they came into musth, they abandoned their bull areas and their male companions and moved off in search of receptive females. During their musth periods the time males spent feeding and resting declined markedly, and instead they occupied themselves with walking around interacting aggressively with other males and pursuing females. The difference in behavior was particularly pronounced in the oldest males.

On graph paper I began to plot the association patterns of my focal males against time, marking whether they were alone, with other males or with females, and whether they were in musth. The results were fascinating. When not in musth, the older males spent their time with other males or alone; as musth began, they kept to themselves for a brief period and then for several months were either in the company of females or alone, moving from one female group to another. As musth subsided they resumed a brief period alone, and as musth ended they returned to the company of other males. All of the older males showed a similar pattern, except that each male's musth period occurred at a different time of the year. For example, Dionysius's musth period was in February, March, and April, Iain's was in March, April, May, and June, while Bad Bull's was in June, July, and August.

The pattern exhibited by the younger focal males was quite different. They, too, spent a proportion of the year with males, but occasionally they interrupted this stay for a visit with female groups. Similarly, the months they spent with females were interrupted from time to time for a visit with males. Early in my study I never observed the younger males in musth during their sexually active time with females, but then an interesting thing began to happen: Some of them began to come into musth, but only for a day or two, after which they would drop out of musth again. An even more fascinating observation was that these younger males tended to be in musth when none of the older musth males was present. Could the older males prevent the younger males from coming into musth through sheer intimidation?

Based on my observations and the patterns that had emerged on my graph, I was able to define three distinct sexual states: sexually active musth, sexually active nonmusth, and sexually inactive nonmusth. But, I began to wonder, how did these different states relate to the males' testosterone levels? The hormone testosterone controls sexual and aggressive behavior in males. Based on the very aggressive nature of my sexually active musth males, I would have predicted that their testosterone levels would be significantly higher than those of the sexually inactive nonmusth males. But what about the males who were sexually active but not in musth? To answer that question, I had to find some way to measure the levels of testosterone of males in each sexual state.

In 1980 the standard technique for monitoring hormone levels in both captive and wild animals involved immobilization and the collection of serum samples, but I did not want to interfere with my males in this way. I had, however, heard that testosterone had recently been isolated from the urine of captive animals at the San Diego Zoo by veterinarians Bill Lasley and Lonnie Kasman. Collecting urine would be less intrusive and dangerous for my elephants than immobilization, though perhaps not less dangerous for me. I contacted Lasley and Kasman, who agreed to analyze some urine samples. If I succeeded in obtaining them, it would be the first time that testosterone levels were extracted from urine collected from a free-ranging mammal.

For statistical reasons I needed to gather preferably ten urine samples from males in each of the three sexual states, and I needed to collect urine from the same male in at least two different states. Under normal circumstances an adult male elephant urinates five to ten liters of urine some fifteen to twenty times a day. It takes a minute or so for that amount of urine to soak into the soil, which was all the time I had to drive up to the bull in question and persuade him to move over a few steps so that I could get out of my car and aspirate the urine from the ground with a twenty-milliliter syringe.

This procedure proved to be surprisingly easy, at least until I tried to collect urine from males in musth. Musth males don't urinate, they dribble, and the only time they dribble in any useful quantity is when they are annoyed. The other problem was that musth males are usually not standing still when they dribble in quantity, except when they are listening to something in the distance that irritates them, like the rumble of another musth male, an airplane passing overhead, or a truck driving by. The usual sequence of events was that I would spot a musth male dribbling at a sufficient rate and see a tiny puddle of urine seeping rapidly into the ground. There was no other option but to drive up and try to urge the male to move. Inevitably this would anger the male even more, and although the high rate or dribbling would continue, he would now be towering over my car, ear-waving and musth-rumbling. By the time he had strode off far enough for me to risk getting out of the car with my syringe, all that remained was a strong-smelling wet spot on the ground. Eventually, however, despite a number of close calls, I did manage to collect enough samples from males in musth.

Once analyzed, the testosterone levels supported what the behavior of the males had indicated: that musth males had extremely high levels of testosterone, sexually active nonmusth males had moderately high levels, and sexually inactive nonmusth males had very low levels of testosterone.

By the end of 1980 the secret life of Amboseli's male elephants was beginning to come to light. But there were still so many puzzles. Why did male elephants come into this "intoxicated" musth state? If sexually active but nonmusth males can mate, what advantage did their aggressive behavior confer? How did musth, body size, and condition interact to affect dominance rank? Were musth males better able to compete with nonmusth males for matings? Was an estrous female more likely to show preference for a male in musth?

I was fortunate in my pursuit of answers to these questions. The 1981 long rains were good, and I couldn't have chosen a better time for my study. Wildflowers bloomed, the grass was long and green, and the elephants got into aggregations of up to several hundred individuals. Females came into estrus two or three at a time, each pursued by scores of nonmusth and several musth males. May of 1981 saw a record twenty-four females in estrus and thirteen males in musth.

CHAPTER ➤10

Dangerous Encounters

No animal in the world is as dangerous
as an elephant in musth.
—Charles Darwin, *Origin of the Species,* 1859

During the first few months of my study, I often erred in judgment as I followed a musth male too closely or into an area of thick
bush where there were limited avenues of escape. While my survival depended on being able to interpret when their threats were
serious and when they were bluffing, at the time I was still a relative
novice at discerning the moods of elephants, and on many occasions
my own miscalculations brought me dangerously close to death.

One such occasion occurred when I followed Dionysius through
a maze of palm trees deep in the heart of *Oltukai Orok.* He was in
full musth and guarding an estrous female who was a part of a group
of over fifty elephants. Agamemnon was one of them, and as he
was also in musth he, too, wanted the female. Around and around
the palms the two males went and, though I knew that I was court-
ing danger, I trailed closely behind them, for I needed the data on
male-male aggression, on guarding behavior, on mating success, and
on female choice. This was a perfect opportunity, and the chase
was exciting.

We continued to circle the grove of palms, Dionysius trying to guard the female, Agamemnon doing his best to get her away from him, and me trying to record everything that happened. The high pitch of my two-stroke Suzuki engine whined as I maneuvered this way and that. I was getting in Dionysius's way, and I could see by his repeated folded ear threats and his head-high posture that he was becoming irritated. The elephants picked up their pace and suddenly the female doubled back. Agamemnon found her first and mounted her, Dionysius charged through the palms and knocked him off, and Agamemnon bellowed in fear. Then they were off again.

I drove with my right hand and held a Dictaphone in my left. After following them through a narrow gap between the palms, we entered a beautiful little glade that was bordered by swamp on one side and a wall of palms on the other. Dionysius disappeared around a tall palm and slipped back out of the glade again. There were two exits, the passage I had come in by and the one through which Dionysius had gone out. I followed closely behind him until he suddenly swung around and came for me, his ears folded under, his head and tusks down. This was no longer a threat, he meant to tusk the car! I yanked the steering wheel hard to the left, floored the accelerator, and tried to escape through the other exit, only to find that he had circled around and was now charging at me from that side. In a dead panic I drove back to where I had been seconds before only to find him once again coming at me, head down. He was behaving as if he thought that I, too, wanted his female. Back and forth we went, and each time he blocked my escape. I was about to be killed! Perhaps I should abandon the car and try to cross the swamp through the thick papyrus, I thought. I could just see the camp on the other side. Maybe Dionysius would be content with destroying the car and would forget about tracking me. I talked aloud to myself, sure that this might be the end.

I decided to try the way I had entered one more time before jumping into the swamp. "Okay, Joyce," I said to myself, "now go, fast, it's your last chance!" I drove through the gap as he came for me, head down, ears folded, with only a few feet to spare. Ahead a wall of elephants blocked the way, including Agamemnon and the estrous female. "Just go!" I shouted as I floored the accelerator and drove straight at them. With a thunderous roar, the huge gray bodies parted, dust flying everywhere. Although I could still see Dionysius's wide tusks in the rearview mirror, it looked as if I would escape, but I was still driving for my life. As I rounded the palms I noticed Cynthia driving slowly toward me in the distance. She stopped her car, as she always did, to hear my elephant news, and I burst into tears. It was then I discovered that my Dictophone was still running, leaving me with a permanent record of my folly.

In late May Bad Bull came into musth, and the reign of terror started in earnest. I knew he was the one elephant who would not hesitate to tusk the car, and for some reason, he particularly hated research vehicles. After Cynthia's experience in 1977, when he had tracked her down to the swamp edge, Bad Bull's insidious style had become even more refined. He would stand perpendicular to the car, pretending that he was doing something else, but all the while watching out of the corner of his small, evil-looking eyes. Then, just when we were lulled into believing that all was safe, he would pivot and come for us. His other tactic was to pass the car in what we called the "parallel walk" and then spin around and attack. Such Bad Bull tales had continued to accumulate, and I couldn't help but recall some of my own experiences with him as I tried to summon the courage to get close enough to collect the data I needed.

Once, on the plains south of *Longinye,* on the edge of the dying grove of acacias, I had a terrifying encounter with Bad Bull. I felt quite safe there because I knew that my car could move faster than Bad Bull could and there was nothing out on the plains to slow me down. I was using the Jolly Green Giant, an old Toyota Land Cruiser belonging to the African Wildlife Foundation, and when Bad Bull charged, I turned the key in the ignition and slammed it into first gear. To my horror I found myself shooting backward across the plains with Bad Bull bearing down on me. In my terror,

I had forgotten that the Jolly Green Giant was one of those old-style Toyotas that had only three forward gears, and reverse was in the position that one normally expects to find first. As Bad Bull's tusks came closer and closer to the front windshield, I remember recalling Iain Douglas-Hamilton words: "You can never escape from an elephant in reverse." I had to get the car into first gear. My only hope was that the sudden change in the car's direction would startle him enough to give me time to escape. I jammed on the brakes and threw the car into first just as the elephant lowered his tusks toward me. Then, as the car started moving in his direction, Bad Bull paused, giving me the split second I needed to veer past him, and I roared off across the open plains with him charging after me. I had made it, but I was shaken to the core.

After close calls like this Cynthia and I both endeavored to keep a very safe distance from Bad Bull. Keith Lindsay, who was then studying elephant-habitat interactions for his master's thesis, once tried to stand up to him; to this day I can't decide whether this challenge was a demonstration of Keith's manhood or just plain stupidity. We were both observing a big aggregation of elephants on the edge of the tortilis woodlands. I was watching Bad Bull in the Jolly Green Giant from a safe distance of at least 100 meters. Keith was in the same group in a little white Suzuki with his assistant David Maitumo. Bad Bull had moved in close to their vehicle and was starting to do the parallel walk. I knew from the evil look in his eye that he was about to attack, but I was too far away to warn Keith and David of the impending danger. Suddenly Bad Bull spun ninety degrees and went for them, his head down, ready to tusk their little car. As Bad Bull neared their vehicle, Keith and David did a most extraordinary thing—they opened both doors and waved them frantically in and out.

Bad Bull stopped short several meters from the car, lifted his head high, folded his ears, sidestepped, and kicked dust in the air. Then, using a standard musth male fighting tactic, he pretended to walk away. Keith started up the car and tore off in the opposite direction as Bad Bull spun around in close pursuit. Once it was safe, I drove over to Keith and asked him what on earth he thought he was doing. He explained that they had been flapping their ears to threaten Bad Bull! I don't believe he ever tried that strategy again.

Now I found myself again in the presence of Bad Bull, warily doing a half-hour follow on him. He was devouring an acacia tree near camp on the edge of *Oltukai Orok* and had turned so that he could watch me out of the corner of his eye. I had parked my little green Suzuki a good distance away from him and at an angle that allowed me a clear route out. The timing of the two-stroke engine went out frequently, and the spark plugs oiled up at least once every day. Consequently, the car had a nasty habit of spluttering along and then dying, and I had narrowly escaped several other musth males with the engine firing on only two plugs. I knew that Bad Bull would attack before my observation period was over, and I knew that I had to be ready. But, as always, he caught me by surprise. I've never been a very clear thinker under pressure and, when he attacked, I panicked, put the car in second gear, and stalled. By the time I had started the car again and got it moving, I could see his tusks in the rearview mirror only centimeters from the back window. Again I floored the accelerator, and the car gave a long high-pitched whine as it bumped across the plain. Mature male elephants can move at a good forty kilometers per hour, and the two-stroke Suzuki couldn't do much more than that even at the best of times. I'll never know quite how I escaped.

Several weeks later I was talking to Cynthia in her tent late one afternoon when I looked up and saw Bad Bull following a group of females into camp. By that time Cynthia and I were both being terrorized by him regularly. He seemed able to identify both of our vehicles, whether by sound or smell, and would come for us from a good four hundred meters away. Now he was heading straight for the tent, and we felt certain that if we stayed where we were he would smell us, attack the tent, and gore us both to death. We decided that our only hope was to sneak out of the tent and run for the car, which at least offered a thicker layer of protection than canvas. As we ran from the tent to the car, he spotted us, spun on his great feet, and charged. Hiding behind the car, we held our breath as he stood tall. Luck was with us, for he had an estrous female to guard, and he abandoned his attack, leaving us breathless, hearts pounding and hands shaking.

During Bad Bull's 1980 musth period I became so anxious that I frequently dreamed about him. My most common dream involved being chased across the plains. He could run faster than my car and he was gaining on me; I had to reach the *Enkongo Narok* swamp causeway before he did. I reached its entrance, but the causeway had washed away; Bad Bull's tusks appeared in the rear-view mirror . . . Then I always woke up.

In another of the more vivid dreams I was involved in a long conversation with Bad Bull during a period when he was out of musth. I had summoned up all of my courage and gone to look for him. I found him feeding by the edge of the swamp at Place of the Bulls and called to him. "Bad Bull, I need to talk to you." He was surprisingly obliging and said, "Yes, what can I do for you?" I had expected to have difficulties persuading him to talk and so I had brought a *gunia,* or a burlap sack, full of malted milk balls as an incentive. We sat side by side, our feet dangling in the crystal-clear water, eating our way through the malted milk balls. I carefully broached the subject: "Bad Bull, do you know that during your musth period you've almost killed me?" "Musth?" he asked. "What is musth?" I was surprised that he didn't know about it. "Musth is that time of the year when you secrete from your temporal glands, dribble urine, and your penis turns green," I explained. "Dribble urine and my penis turns green? How fascinating. I must say, I wasn't aware," he remarked. I pressed on: "Well, during that time you are very aggressive, and you and the other bulls have almost killed me several times. I just wanted to let you know that I am not trying to compete with you or steal your females, I am just doing a study of your behavior. I wondered whether you would be so kind as to tell the other males what I am doing and ask them to leave me alone." Bad Bull agreed to speak to them and assured me that I would not have any more trouble.

My study continued peacefully until one day I learned that the scientific community had decided that my results were not valid because my focal animals knew what I was studying and would, as a result, behave differently in my presence.

Sometime in the middle of Bad Bull's 1980 musth period I received a letter from Robert Hinde, who said that he had heard that I was being seriously charged by musth males and that he thought I was taking unnecessary risks. He was right, I realized, and as a result of his warning I made some new rules for myself: I was not to go within fifty meters of a musth male, and I was not to follow a musth male into thick bush.

With time I learned to read and understand the different signals of the musth males. I discovered that turning off the car engine each time I was threatened taught them that I respected their dominant position and that I was not competing for their females. Gradually we came to trust one another, and I felt comfortable enough to abandon my rules. Eventually my greetings and interactions with each musth male became ritualized: RBG musth-rumbled at me each time I moved the car to catch up with him; Agamemnon stood with his forelegs against the front bumper of the car, touched the windshield with his long asymmetrical tusks, and then swung them back and forth over the roof of the car, looking down at me through his long eyelashes; Alfred loomed over the front of the car, musth-rumbling and "drawing pictures" with his trunk in the dust on the hood; with Sleepy I played a game of musth male chases researcher across the open plains; Dionysius and Iain stood calmly next to my car as friends. To this day there is only one musth male I never allow near the car: Bad Bull.

CHAPTER ➤ 11

My Amboseli Home

I CONTENT MYSELF WITH LIVING AMONG THEM. I LIKE THEM. I LIKE
LOOKING AT THEM, LISTENING TO THEM, WATCHING THEM ON
THE HORIZON. TO TELL YOU THE TRUTH, I'D GIVE ANYTHING
TO BECOME AN ELEPHANT MYSELF.
—Romain Gary, *The Roots of Heaven*, 1958

MY PROJECT WAS PROGRESSING well, and life in the Elephant
Camp was also to my liking. The mood in camp inevitably changed
with the cast of characters. When I had returned to Amboseli in
January 1980, I found Pili packing up. Pili was our nickname for
Phyllis Lee, after *pili pili*, chili peppers, for her often fiery temperament and sharp tongue, which, much to my chagrin, she used on
me regularly. Cynthia consoled me by accurately describing Pili as
a marshmallow, soft and gooey on the inside, and once I started
calling her bluff we established a close friendship that lasted through
the many ups and downs of thesis writing, relationships, and Amboseli's isolation. Having completed the fieldwork for her study of
development and maternal investment in vervet monkeys, Pili was
on her way back to Cambridge where she, too, was to write her
dissertation under the supervision of Robert Hinde.

Keith Lindsay had recently departed, having completed a study
of elephant feeding behavior for his master's at the University of
British Columbia. I didn't know Keith well at that stage, having

overlapped with him only for short periods, but he had a wonderfully wicked sense of humor, and I missed his quick wit and amusing stories. I got to know Keith better when I returned to Cambridge to write my dissertation and later when we were again in Amboseli together. Over the years we became very close, and Keith was a source of tremendous support and warm friendship.

Both Keith and Pili had plans to come back to Amboseli, Keith to expand on the study he had done and Pili to switch from vervets to elephants. But once Pili left in late January, the camp consisted of just Cynthia, Masaku, the cook, and me. Cynthia and I spent the long days watching elephants and the evenings recounting elephant gossip by candlelight: who was in estrus, who was in musth, who had mated and with whom. It was like reading a long novel that you don't want to end. I became so engrossed in the elephants' lives that on Sundays, when it was my duty to guard the camp, I used to will them to take the day off, too, so that I wouldn't miss any important events in their lives.

There can be few experiences as rewarding as living in the middle of nowhere with one's closest friend, and when after July Cynthia Jensen came to stay in camp, life in Amboseli was close to perfect. At college we had taken the same courses, played lacrosse on the same team, and run for miles together each day. The day-to-day challenges and isolation of life in the bush can make or break a friendship, and through our experiences together Cynthia and I became bonded for life.

Cyn, as we called her, to distinguish two Cynthias, had come to carry out an ecological study under David Western's direction, analyzing the changing quality of grass in and out of the Amboseli basin and its effect on the migratory patterns of wildebeests. It was her first time in Africa, and the day she arrived we drove across the dry lake singing Bonnie Raitt tunes into the wind as the dying rays of the sun turned the dust of the lakebed to pink and the mirages faded in the dusk. Tall and blond with high cheekbones and blue eyes, Cyn had classical Danish looks and a warm and generous nature. She excelled in Ecology at both Smith and Harvard and later at Cornell—where she undertook her Ph.D—and yet was never convinced of her many talents.

By December the atmosphere in camp had changed again with the arrival of Robert Seyfarth and Dorothy Cheney and their student, Sandy Andelman. Robert and Dorothy studied vervet monkey behavior and cognition. They were an intellectually stimulating and entertaining couple to have in camp, and six people seated around a three-foot-diameter dining table made dinner a lively event.

No matter who happened to be in camp, we all woke at dawn when the trees and the palms stood out as black silhouettes against the salmon-pink horizon and the mountain rose in the south a deep mauve against lavender. Dawn was a time for quiet reflection, and I often contemplated my life with deep satisfaction: I was living in a tent beneath the African sky, studying the behavior of the animals I loved. If I had any sadness it was a wish that my father could have been there with me and that my mother and siblings would come to visit.

I had chosen a perfect spot for my tent in the shade of a grove of palms with a view of Kilimanjaro. My small home was furnished only with essentials: a folding camp table served as a desk, a small cupboard stored my clothes, and I slept on a *jua kali* bed with springs fashioned from one-inch strips of old inner tube crisscrossed from side to side. It was uncomfortable, but one step up from another variety made from old sacking and ropes, known as *teremuka tukase,* "get off so that we can tighten the bed." I had decorated my quarters by hanging some beaded Maasai necklaces from the ceiling, laying some colorful coastal *mikeka,* or woven mats, on the floor, and covering my bed with a Somali *kikoi.* Over the tent I built a palm frond (*makuti*) shelter to keep out the rain and to protect the canvas from the relentless African sun.

Although we had no alarm clocks, no meetings to make on time, there was a rhythm to life in camp and our days were punctuated by certain ritualized events. Two of them took place immediately after sunrise when my thoughts and the tranquility of an African dawn were shattered first by the loud honking of Egyptian geese and then by Masaku's singing as he brought tea to my tent:

> *Watu wa thambi, wanalia*
> *Kwa sababu wanatupwa motoni . . .*

> The sinners are crying
> because they are being thrown into the fire . . .

Masaku, our cook, was the strongest personality the camp has ever known, and those of us who shared the camp with this remarkable individual will never forget him. He was a thin, wiry little man with only one eye, and he was prone to erratic mood swings. Masaku was no longer young, having retired as a railway laborer perhaps a decade before he joined us in Amboseli. He had sung his hymn to me daily ever since the first night I shared my tent with a man.

Masaku liked to think that he had known me for a very long time. When he was in one of his highs, he would claim to visitors that we had been aquainted when *matiti ilikua na simama,* "when my breasts had stood up," and he demonstrated what they looked like by bouncing on his toes and pressing his fists to his chest, his thumbs pointing up and out. Since I was nineteen when we first met and not more than twenty-three when these stories began, his tales did not amuse me. My displeasure only encouraged him further and he would laugh aloud as he stuck his thumbs in the air.

Lying in bed sipping the milky sweetness of Masaku's tea, I listened to the growing chorus of morning sounds: yellow-vented bulbuls twittering in the palms, a pied wagtail's clear song as it paced back and forth on my *makuti* roof, zebras calling to one another on the plains. My morning reverie was usually interrupted by another dawn event: the arrival of the Ts, which was heralded by nervous laughter from the direction of Campsite 18. The Ts were members of the *Oltukai Orok* clan, and the occupancy of the public campground determined their daily routine. Tuskless, Tonie,

Tilly, and the boys, Taabu and Teddy, arrived at the North Clearing just after sunrise, from where they began their morning raids. Lying in bed, I could follow their successes and failures by listening to the reactions of campers.

Typically, a raid started with sounds of human surprise and amusement, and then, as Tuskless and company made clear their intentions, the laughter turned rapidly to screams of panic and hysteria. Then came the banging of pots and pans, the hooting of horns, and the revving of car engines. The noise eventually died down once Tuskless had devoured everything she desired, only to start afresh at the next camp along her route. Once I had watched her eat one bag of maize meal after another, stuffing them into her mouth whole, and then turning to a box of tomatoes, scooping them up five at a time. Having had their fill, the elephants would meander off to one of the lodges to feed at the rubbish pit, timing their return to the campsites when supper was on the table. Although I frequently rescued campers from elephant raids, I was rarely inclined to do so at dawn. Chasing the elephants away meant getting into my car and driving at them full speed, horn blaring. They had yet to call my bluff.

Many campers made the mistake of leaving food in their tents when they went off for an early-morning game drive. During the peak raiding period, Tuskless and her friends and relations destroyed several tents each day. Eventually, after advice from us elephant researchers, a sign was erected warning campers of the elephant raiders and urging them to store all food in the trunk of their cars. The elephants, however, were not deterred in the least and quickly solved the problem by prying open the trunks with their tusks. Eventually the public campsite was moved south of Observation Hill, well outside Tuskless's home area, and only our elephant camp remained.

I usually tumbled out of bed at six-thirty; threw on a pair of moderately clean shorts, a T-shirt, and a pair of sandals, quickly splashed some water on my face from the jerry can outside my tent, and made for the kitchen. Then I braced myself for another camp ritual. As I prepared a Thermos of tea to take to the field Masaku fixed me with his one eye and a deep frown, tapping a knife rhythmically against the side of the table as he waited for my order. *"Masaku, leo nigependa mayai sklumf na tosti,"* I would like scrambled eggs on toast, I told him. The Akamba tribe, of which Masaku was a member, have difficulty pronouncing the letters b, p, and r. Mayai *sklumf* was scrambled eggs, *fly* was fried, *foilo* was boiled, *focho* was poached. I was Joycie Fool and Cynthia Jensen was Sin-Sing Jackson. Masaku claimed that if women didn't eat eggs for breakfast, they would fall over.

After breakfast I put my Thermos of tea in my jeep and checked that everything I needed was in the car: binoculars, camera case, sunglasses, bull box, cow cards, notebooks, data sheets, cool box, and urine vials. It was exasperating to have to come back to camp from the field to pick up some forgotten item. I didn't bother to take food out with me, because I had a tendency to eat it all by nine o'clock.

When I returned to camp in the afternoon I usually found Masaku sitting on his stool outside the kitchen chewing on a grass stem and frowning at the world—another camp ritual. *"Habari Masaku?"* How are you?, I asked. *"Kama jana."* The same as yesterday, he said with a grunt, stretching his arm out and pointing to the pains in his joints. Everything ached, life was not good. I don't know why I bothered to pose the question, except that it was worse to walk past him to the kitchen knowing that he was glowering at my back. When Masaku was on a high he skipped to the *choo,* the outhouse, and threatened to paint everything in the camp green, including the trees. With a glint in his eye, he would produce sausages standing erect in the mashed potatoes for dinner. He raced around as if possessed, doing everything you had requested him to do and more. These moods lasted for days, sometimes weeks, until we almost yearned for the peace and quiet of his lows. Then, as suddenly as a high had begun, a low set in and he returned to his stool outside the kitchen and glared at everyone with his one eye.

We ate a late lunch in the dining room tent, carrying food on a tray from the kitchen and glaring at vervet monkeys who tried to intimidate us with their white eyelids and blue balls. We learned from experience and from our vervet colleagues, Robert and Dorothy, that shouting at them only indicated our subordinate status and increased the likelihood that one would snatch some food off the tray. Silent scowling with raised eyebrows forced them to look away and scratch themselves nervously.

During the heat of the afternoon I worked at my desk collating and analyzing data. A light breeze blew through the tent, and the palm fronds rustled softly against one another. The *Oltukai Orok* clan was often around to take my mind off the monotony of data entry. We were usually aware of their arrival long before they appeared: a deep vibrating rumble or the rasping sound of palm fronds brushing against rough skin as the elephants moved along their hidden trails in the *Phoenix reclinata*.

The families of Tuskless, Tania, Remedios, Xenia, Estella, and Echo were our most frequent visitors, but other families came and went, too. The elephants followed a regular daily pattern, either arriving in the North Clearing from Crossing Corner and moving slowly through the camp to the South Clearing, or arriving in the south from the center of *Oltukai Orok* and moving through to the north. While they often simply passed through the camp on their way somewhere else, the Ts occasionally spent the whole day with us, feeding around the tents. After Tuskless knocked down our kitchen twice, eating or destroying almost everything in it, Masaku collected wine and beer bottles, which he threw at any elephant who ventured too close. While Tuskless eventually learned to be on her best behavior with us, a bottle only infuriated the young males Taabu and Teddy, who chased us around the camp if we so much as lifted our hands toward them. They often pretended to feed at the perimeter of the camp, slowly edging closer and closer to the kitchen, until one of us had to get in the car and chase them away. As Teddy and Taabu got older, larger, and more confident, controlling them became increasingly difficult. They knocked over the clothesline, walked off with basins and hose pipes, and stepped on our water jerry cans. Then one day Taabu just disappeared, and Teddy, having reached his late teens, decided that he was more interested in females than the food in our kitchen and went off to become a normal young male. Our problem was over. Tuskless still visited us frequently, but continued to respect the camp rules. We eventually threw the wine bottles away and allowed her to wander freely among the tents.

As I sat in my tent entering data into my blue analysis book, I often paused to look out over the South Clearing and up to Kilimanjaro. No matter how many times I studied the mountain from camp it never looked quite the same, changing as it did with the time of day and the time of year.

During the long dry season a gloomy haze descended upon Amboseli. Everything was drained of color and vitality, and days would pass without a glimpse of the mountain. At that time of year our lives, too, seemed to lack definition, and we suffered from a dull feeling we called the "Amboseli fuzzies." After the rains began, the "fuzzies" vanished and a sense of well-being, a lightness of heart filled our lives. With the rain all of the dust was washed from the sky, and it seemed as if we could see every tree on the lower slopes of Kilimanjaro.

On those clear days it was the wheat farm across the border in Tanzania, near the small crater of *Legumishera,* that always attracted my attention. During my early years in Amboseli my attention may simply have been caught by the striking contrast that the wheat presented to its surroundings—a big, square splash of uniform color, yellow ocher in the late dry season, emerald green after the rains— against the mottled purples, blues, and greens of the forests and savannah on the lower slopes. Later it was the elephant trail passing near the wheat farm that drew my eye.

To the east of *Legumishera* and the wheat farm was a gorge, and along its ridge ran the elephant trail. With my binoculars I could just make out the dark, twisting line of the gorge, which I knew began in the forest just east of the wheat field, somewhere near the sawmill. We spoke of the sawmill as if we knew it well, when, in fact, our experience of it was only a dot on the map or a cluster of small buildings under the wing of a plane.

It was the trail along the gorge that brought the Kilimanjaro elephants to Amboseli, some said in search of salt, others said to escape the slippery mountain slopes during the rains. As I mused about the trail and its elephants, I preferred to think that they came in search of mates, a plausible explanation, in fact, since they always arrived during the breeding season, when our biggest males were in musth.

The Maasai call the trail after an elder, Ole Sarere, who loved elephants. He was said to have sat by the edge of the trail just so that he could watch them go by, and he often followed their footprints so that he could catch a glimpse of them disappearing up the trail. One day he found an elephant's amniotic sac and placenta on the trail, which the Maasai believe brings good luck and great wealth. Ole Sarere decided to build his *enkang,* or settlement, on that very spot by the elephant trail, and he died a very rich man.

Our Kilimanjaro visitors were families of long-legged, ugly elephants, generations of prune heads: small, dark creatures with sloping, wrinkled foreheads; thin upwardly curving tusks; and small triangular ears with parallel venation patterns. The Kilimanjaro elephants typically had no hairs on their tails, and I could think of no better explanation for this condition than that they must have stepped on one another's tails, breaking off the hairs, as they followed one another down the steep mountain trails.

In the early years they ran away at our approach with their heads and tusks high, their tails up, and the whites of their eyes showing. I wondered what ordeals they must have been through up on their mountain to be so terrified of humans. Many years later, when they had begun to calm down, it was explained to me by the *Olaiguenani,* the chief, Ole Musa, that the Kilimanjaro elephants had traditionally been fiercer, tougher, and more terrifying that their slower, lazier Amboseli cousins. The Maasai warriors always knew that they could catch up with a family of Amboseli elephants, but the Kilimanjaro elephants could disappear into the thin mountain air. On the trail of Kilimanjaro elephants you could put your finger in a pile of fresh dung and find it still warm, you could find a puddle of urine that had not yet seeped into the soil, but try as you might, you could not catch up with the family that had left them. The Maasai believe that the Kilimanjaro elephants obtain this power by eating the bark from the *Olkonyil* tree. The warriors make a soup from the tree's bark, and drinking it puts them in a fighting frenzy, giving them courage to defeat the enemy.

Most afternoons also involved some form of camp chore: fetching water, collecting firewood, or buying vegetables or beer in *Ol Tukai*. Whenever we ran out of fresh vegetables we could purchase potatoes, tomatoes, onions, and cabbages from the *wanawake wa mboga,* the vegetable ladies, there. After we couldn't stand another meal of cabbage, it was time for a trip to Nairobi. Though we cursed the safari and all the headaches of the big city, these excursions also were occasions to socialize and we planned them carefully to coincide with parties organized by friends. They also provided an opportunity to get together with wildlife researchers from other parts of the country, tell stories about our animals, our adventures and narrow escapes, and our dismal love lives.

In the evening, when the sun descended below the palms, the acacias shone golden, the base of the mountain turned slate blue, and the dying rays of the sun cast a splash of pink on the western wall of the Kibo glacier. Masaku heated water in the drum and if he was in a good mood he filled the *debe* for me. I dropped my dusty clothes on the floor of my tent and wrapped one of my *khangas* around my brown body. At the end of the day it was hard to tell whether my coloring was due to the sun or the dust. I slipped on my old flip-flops, noticing that dust had turned to mud by the perspiration between my toes.

Masaku's daily warning to me that the water was hot was *"Usichomeka kama kuku,"*—Don't get burnt like a chicken—presumably meaning the water was hot enough to remove feathers. Water was precious, and I turned off the tap whenever it wasn't essential. I never used more than ten liters and as I walked back to my tent

Masaku always shouted after me, *"Wewe, Joycie, unaoga kama ndege,"* You bathe like a bird.

There was something very sensuous to me about showering in the open air and watching the elephants walk by. I took great pleasure in being wrapped in nothing but a thin *khanga* feeling the cool air lightly caressing my body. Back in my tent I put on a long-sleeved shirt, wrapped another *khanga* around my waist, and joined the others in the dining room tent for a beer.

Over dinner we recounted the day's stories, laughing at some of the absurdities of living the way we did. We walked back to our tents by moonlight calling out our good nights. I always stood outside to brush my teeth, breathing in the night air and enjoying the sense of anonymity that comes with darkness. As my eyes adjusted to the dim light, the moving shapes became familiar animals. Some nights I thought I saw a leopard slinking through the clearing, swishing its tail up over its back, but as I stared into the darkness whatever it had been was gone. Then I switched on my twelve-volt light and stepped inside my little home on the plains.

CHAPTER ➛12

Births and Deaths

IN THE DAILY SEARCH for my males, I stopped to record each group of elephants that I sighted, documenting any important or unusual behavior in my notes. Although the interactions and fights between musth males were always thrilling to watch, I think the most extraordinary elephant behavior I witnessed during my eighteen-month study was the birth and death of two female elephants, and the reaction of another to the birth of her stillborn infant.

Early one morning I was driving along the southern edge of *Oltukai Orok* when I came upon a group of eighteen elephants. It was Teresia's and Slitear's families, part of the T bond group. I stopped to take some notes and noticed that one of Teresia's granddaughters, a young female named Tallulah, was acting strangely, lowering her hind legs and dropping down on her knees. As I studied her behavior closely, I noticed that a slight bulge had appeared below her tail, that her vulva were swollen, and that she was urinating in a continuous dribble. The bulge under her tail confused me until I realized that it was moving slowly downward.

The dropping down on her knees had been a contraction; Tallulah was about to produce a baby!

As I fumbled with my camera, my hands trembling with excitement, Tallulah reached the edge of *Oltukai Orok* and disappeared from view in a clump of Phoenix palms and regenerating acacias. I found her lying on her side under a palm at 8:38 A.M. Leaving my car, I crept slowly toward her so that I could get a better view. A minute later Tallulah stood up, and I noticed that the bulge had become larger and had descended farther down her birth canal. At 8:41 Tallulah looked directly at me and walked away into a patch of acacias, only to turn around and come back toward me. Two minutes later she walked away again and disappeared into the acacias. I drove quickly to the other side of the clump of trees to find her baby on the ground at 8:44, still contained inside its amniotic sac.

For a minute Tallulah stood calmly and quietly over her newborn, and then gently touched it with her forefoot. The baby kicked its tiny feet in response. Tallulah bent over and, using her tusks, freed the newborn from the sac. Then she began to scrape the earth with her front feet, clearing the vegetation away. Her baby moved again, and Tallulah became more excited, scraping at the ground with renewed vigor. At 8:47 some members of her family arrived, rushing in to surround Tallulah and her newborn, with their heads and ears high, urinating and rumbling loudly. Bending down and using her feet and her trunk, Tallulah tried to get her baby to stand. The other females gathered around, continuing to scream, rumble, and trumpet, as temporal gland secretions streamed down the sides of their faces. Tallulah's high-ranking aunt, Slitear, now made her entrance, backing into the group amid more urinating, rumbling, and trumpeting. Tallulah continued to scrape the ground, tried to lift her baby, scraped the ground again, and then gently touched the infant with her trunk.

From my position, fifteen meters away, the newborn was barely visible through the scores of legs, trunks, and tails. At 8:56 the baby was pushed to its feet but toppled over. Two minutes later it was once more on its feet, only to fall over again. It took the baby more than half an hour of struggling before it finally managed to steady itself on its four legs. New females continued to arrive, some from Tallulah's bond group and others from outside families. The excitement in the group was intense, and several females dropped to their knees and, with their heads held high in the air, flopped their trunks about and gave me a wild look, the whites of their eyes showing. It was as if, in their excitement, they had forgotten that I was not an elephant and wanted to share their emotion with me. Another young female picked up the amniotic sac with her trunk, waved it in the air, and then tossed it over her head.

Finally, at 9:21, forty minutes after its birth, the baby took its first steps. The group began to make its way slowly along the edge of the palms, leaving the birth site, an area eight meters in diameter, trampled beyond recognition. At 9:47 the newborn passed its meconium, or fetal stool, and at 10:00 it suckled from its mother. The new baby was a female and Cynthia later named her Tao.

Several months later, early one morning in the late dry season,

Cyn had driven from camp to *Ol Tukai* and noticed an elephant moving in a strange manner, walking and dropping down on its knees, out on the plains beyond the palms. When I returned from the field later that morning I, too, noticed the single elephant, and I thought it strange for it to be standing alone on the plains in the heat of the day. When Cyn returned to camp just before noon, she said that she thought the elephant we had seen on the plains had a dead baby at her feet.

Cyn and I returned to the site together and as we approached the elephant we realized that it was Tonie from Tuskless's family. She was still having contractions, and blood was dripping from her vulva, but her behavior was subdued and so unlike that of Tallulah's. She stood quietly, her head and ears hanging forward, playing slowly, gently with the afterbirth with her trunk. The newborn at her feet was dry, and she repeatedly nudged it gently with her feet and finally rolled it over several times. Two vultures waited nearby. Tonie stayed out on the barren plains with her dead baby for the rest of the day and through the long night.

The following morning Cyn and I left the camp on foot and walked to the edge of the palms from where we could see Tonie still watching over her stillborn infant. Fifteen vultures and a jackal hovered around her; she charged and they scattered for a few seconds, only to return. Tonie placed herself between her baby and the scavengers, and, facing them, she gently nudged the body with her hind leg. As I watched Tonie's vigil over her dead newborn, I got my first very strong feeling that elephants grieve. I will never forget the expression on her face, her eyes, her mouth, the way she carried her ears, her head, and her body. Every part of her spelled grief.

By now Tonie had been standing out on the bare plains without food or water for over twenty-four hours. Cyn and I walked back to camp, found a jerry can, and filled it with water. Interacting with one's study animals, let alone providing water to elephants in national parks, was not really proper behavior for a scientist, but under the circumstances I didn't care.

As we drove toward Tonie she charged, and I stopped the car.

I placed a basin on the ground, poured the water into it, and then drove away. She lifted her trunk toward the water and walked immediately toward it, pausing only once. She drank quickly, emptying the basin in two trunkfuls. I returned to fill the basin again as she stood close by.

Later that morning Cyn and I returned to Tonie with another two containers of water. As she saw me put the basin on the ground she walked over and stood by the car. I held the twenty-liter can on my lap and, with one leg on the ground, I poured water into the basin. Tonie drank while I poured the water onto her trunk, her tusks no more than ten centimeters from my head. After she had emptied both cans, she reached through the door of my car and twice touched my arm with her trunk.

In the early afternoon I returned once again with more water. She emptied the first container and then waited patiently while I banged around trying to get the second can from the back of the car. She drank most of this second jerry can, then used the last bit of water to splash herself. In all Tonie drank ninety liters of water. After she had finished splashing, she again reached inside the car and touched me gently on my chest and arm.

The following morning we found Tonie still on her vigil, attempting to chase away the ever-closer vultures. Later that day she had gone, and all that remained on the plains was a few vultures and scattered bones.

In early July 1981, a few days before leaving Amboseli for the U.S., I spent the morning with Cyn and a Maasai companion, Meloimyiet, watching the elephants in *Olodo Are*. Meloimyiet had for the last six months accompanied me to observe the elephants each day. He had a keen eye and had developed a deep respect for the elephants. My heart that morning was heavy. I had already sold my car and my tent, and I was about to say good-bye to my friends, my elephant companions, and a way of life.

We found a group coming in through the *Acacia tortilis*. Among them was Penelope's family, heading to *Longinye* swamp to spend the heat of the day. It was part of their daily routine at this time of the year, when the long grass under the umbrella-shaped trees had already begun to turn brown. The group was in a playful mood that morning and my spirits were lifted. Despite the fact that the rains had long since ended and the grass on the plains was parched and brown, babies played energetic games under their mother's feet and some of the adult females got down on their knees tusking the ground in apparent delight.

As we watched the scene Meloimyiet suddenly said, "Something is happening over there—elephants are running." Elephants do not normally move quickly, and as I turned to see what was wrong, I caught a glimpse of an adult female toppling over. The elephants standing near her fled as she fell. I drove over in time to find Polly, a female with long, splayed tusks, attempting to pull herself to her feet. Dark bands of temporal gland secretions streamed down the sides of her face, and her eyes were wide and shining, the whites revealing her confusion and pain. She pulled herself halfway to her feet and then slowly started to topple toward us, falling first onto her long splayed right tusk and then, pivoting on it, crashing to the ground beside us. Her four legs curved upward, her great body trembled and stiffed, and she collapsed. We watched from the car saying aloud, "No, Polly, don't die," as if perhaps that might help. She let out a long sigh which reverberated down her trunk, and then there was silence; only a puff of dust remained as evidence of her struggle. I found myself waiting for her to move again, unable to believe that she was dead. "She is finished," Meloimyiet said, as if in answer to my unspoken question. I had watched only one animal die in my life; he had seen many.

The rest of her family had moved on in the direction of the swamp, unaware of her death, and Polly was left alone with only human companions. This was not the way that elephants were supposed to die. I looked around to see whether there were any other elephants nearby who might come to pay their respects to the dead. Approaching us from the tortilis woodlands, still a hundred meters away, were four young males, stragglers at the end of a family group. Following along behind as usual, they had been delayed by boys' games and by all the interesting female smells that had to be investigated. They came with a light step until they saw Polly, when all four stopped still. They lifted their trunks in silent unison and then, one by one, moved toward her. The oldest male, Kasaine, named after Meloimyiet's older brother, took the lead, while the others followed silently behind. Kasaine reached his trunk out toward her body and touched her gently along the length of her back. He then moved to her left tusk and, wrapping his trunk gently around it, he began to pull, trying to lift her. He paused, as if considering what to do next, then slowly lifted his right foot and placed it gently on her trunk and began to roll it backward and forward. He walked around to the other side of her head and, from his new position, he again wrapped his trunk around her tusk and began to pull, trying to lift her. Failing once again, he walked around her body and began to mount her. During the next hour he mounted her over and over again, as if trying to get some response. The other young males continued to touch her trunk and her tusks gently, seeming to search for a reason for her death. Finally they left, walking slowly in the direction the others had taken.

Once the young males had departed, we, too, walked slowly around her body, seeking the cause of her death. Then I noticed the four tiny pricks of fresh blood on her lower left foreleg. Meloimyiet bent over to study the pattern of blood and then quietly remarked in Maasai, *"Olasurai lenkancoi."* The elephant snake, the one with the poison so potent that it can kill an elephant. *"Ile inapanua shingo,"* the one that spreads its neck, a cobra. Was it possible for a cobra to kill an elephant? The Maasai believe so.

Later that morning we left Polly and went to *Ol Tukai* to report her death to the rangers. It was their job to remove her tusks before they fell into someone else's hands. The weight and length of her tusks would be logged in the Ivory Book, then they would be locked in the storeroom until they could be driven to Nairobi, to be kept in the Ivory Room and then exported for trinkets. The rangers left immediately and by midday they had hacked out her tusks with an ax.

The following day we returned to see Polly once more. Her head was a faceless mask of blood. Only her eyes remained, dark and shining under long velvet eyelashes, seeing nothing. Standing over her body were three young males including Kasaine. Like me, he had returned to pay his last respects. Side by side they stood over her, only their breathing breaking the ghastly silence. Their trunks slowly touched the fresh blood where her face and tusks had been only hours before. Their trunks moved over her head, carefully analyzing all of the new scents, fully aware that this butchery was the work of humans. I wondered whether they would see me as different, as someone they could trust, or whether, in their minds, I was just another human.

CHAPTER ➤13

A Sojourn in Cambridge

IN OCTOBER OF 1981 I returned to Cambridge to write my dissertation. Pili had warned me to expect to spend as long writing up my results as I had taken obtaining them in the field. I assumed on that basis that I would be in Cambridge for eighteen months, but I was horrified to find several people at the Subdepartment of Animal Behavior who were in their seventh, eighth, and ninth *years* of thesis writing. I vowed I would not make the same mistake, but watching them agonize over their data, analyzing and reanalyzing and writing and rewriting, filled me with some trepidation.

Having returned from the field in February 1980, Pili herself was in the final stages of her thesis. She was then living in the small village of Dry Drayton, several kilometers beyond the subdepartment, and kindly offered to let me share her cottage for a few months until Keith arrived. (Keith was still at the University of British Columbia finishing up his master's degree and was due to join her early in 1982, when he would start his Cambridge Ph.D. on the feeding ecology of the Amboseli elephants.) Dry Drayton was too far from Cambridge and evening seminars to rely on my bicycle, so I abandoned it for a rusted old Mini Minor, which I purchased for $140. The green paint was peeling off the roof and the wind whistled in through the doors. It had only just scraped by its annual test of roadworthiness and was unlikely to make it through again.

In November Cynthia joined me for a few weeks, and we began entering all of the elephant sightings data from 1972 through 1981 onto a computer. Day after day I sat at the Subdepartment of Animal Behavior computer entering the 3,425 records of bull sightings, while an even greater number of female group sightings were entered on the Cambridge mainframe. With assistance from Pili, who was by then planning her study of elephant calf development, and a computer whiz named Duncan McKinder, programs were written to analyze this enormous quantity of information.

For my dissertation I was particularly interested in determining the frequency of various associations between the elephants, how female group size varied with season, and the size of the groups that estrous females and musth males found themselves in, data that Duncan and Pili both kindly assisted me in extracting. I also had to analyze my focal sample data, collated in my big blue book, and to interpret the dominance and body size matrices.

I was allocated a narrow cubicle in the subdepartment attic where I hibernated for what seemed like months on end. By January I was able to begin putting pen to paper, and soon, with my first chapter in hand, I was nervously knocking on Robert Hinde's door. He summoned me that same afternoon. My confidence wasn't helped by the fact that Robert had a habit of staring everyone in the chest rather than in the eyes. He made a few technical comments, including pointing out to me that "penal" had to do with law enforcement, whereas "penile" had to do with penises, and then simply remarked, "At least you can write." Our meeting was over. I walked out of his office too embarrassed about the penal secretions to absorb the fact that my first chapter was almost complete. I wasn't going to be in England forever!

The winter was cold and damp, and Pili's cottage had no central heating. There was a coal fire in the sitting room, which I found impossible to light, and we took hot water bottles to bed. At one point the water in the toilet bowl froze solid for several days, and I developed chilblains on my toes. As the winter wore on, I decided that I would have to return to Africa before I could face another English winter, and I made plans to return to Amboseli in July for a short holiday. But as I worked my way through chapter after chapter, I realized that if I took a break it would be extremely difficult to pick up the same momentum again: I simply had to complete my dissertation by July. People laughed at me when I told them of my plan.

By April my dissertation was beginning to take shape. I had five data chapters covering the social structure of the Amboseli elephants, the patterns of musth, ecological variation and reproductive patterns, dominance and aggression, and mate competition. As I continued analyzing and writing, interesting patterns continued to emerge. I found that, in general, young males spent up to 70 percent of their time in association with female groups, while the oldest males spent less than 30 percent. It was probably this fact that had led other elephant biologists to assume that older males were reproductively inactive and that it was the young males that were responsible for breeding.

Once males reached their late twenties, the carefree life of the young was over and they entered a highly dynamic social world of volatile sexual state, rank, association, and behavior. An adult male's age and sexual state appeared to determine the structure and size of the groups with which he associated and the types of interactions he had with members of these groups.

Studying my data, I began to piece together a picture of how body size, physical condition, and musth affected dominance relations between males. Between two nonmusth males, dominance was simple: Older, larger males ranked above younger, smaller males. Each male in the population knew his rank relative to every other male, and there were few disagreements except in rare situations where both males were sexually active and one was about to come into musth.

Between two musth males, older, larger males also ranked above younger, smaller males, except when the larger male had been in musth for several months and had lost considerable condition. At such times a smaller musth male could challenge a large musth male. Although rare, this situation could lead to a serious fight that the smaller male often won, causing the larger male to drop out of musth.

In aggressive interactions between a musth male and a nonmusth male, the musth male always won, whether the nonmusth male was larger or smaller, sexually active or inactive. Most cases of musth male versus nonmusth male interactions were between a large musth male and a smaller, sexually active but nonmusth male. But on many occasions I also watched a young musth male enter a bull area and chase one of the highest-ranking but nonmusth males in the population, as if only to give himself a charge of the purest of male elephant satisfaction.

The long-term Amboseli data indicated that the occurrence of estrus and conception in elephants was highly sensitive to rainfall and resource availability. Although estrous females could be observed in any month of the year, the frequency of estrus was significantly higher during and following the wet seasons, when females were in good condition. In Amboseli the highest months of estrus and conception were from February through August.

The seasonality of musth periods reflected the fertility pattern exhibited by females, with most males coming into musth between February and August. What was particularly interesting was who came into musth when. As I pulled together all of the sightings of musth males, I found that the musth periods of larger, older males lasted several months and occurred at a predictable time each year. The musth periods of Amboseli's highest-ranking males, Dionysius, Iain, and Bad Bull, each lasted between three and five months, and, together, these three individuals covered the months when up to 80 percent of estrus occurred. Lower-ranking males were left with two choices: either to come into musth at the same time as these dominant males and attempt to mate with females that they were not able to cover, or to come into musth during the dry season, when there were fewer females likely to come into estrus but also less competition. The middle-ranking males seemed to be divided in their strategy. Those whose sexually active periods overlapped with the high-ranking males came into musth for shorter periods of time, while those whose sexually active periods occurred during the long dry season came into musth for longer. The musth periods of the youngest group of males, those in their twenties and early thirties, were short and sporadic, lasting only a few days or weeks. Musth among these younger males was influenced by the proximity of higher-ranking musth males, who were able to force them out of musth by threatening them.

My data indicated that musth males were more successful than nonmusth males at achieving matings for two reasons: Their large body size and aggressive behavior made them better able to compete for access to estrous females, and the females themselves preferred to mate with males in musth.

Cynthia and I both had been collecting data on the occurrence of estrus and the behavior of estrous females and the males who associated with them. Cynthia had described several behaviors of estrous females in the company of males, which she called *wariness,* the *estrous walk,* the *chase, mating,* and *consortship.* We found that estrus lasted between four and six days. During early estrus the female was chased and mounted by younger males; during mid-estrus she was found, guarded, and mated by a musth male or a series of musth males, and went into consortship with that mate. Consortships are characterized by close proximity, affinitive behavior, and attempts to maintain exclusive copulatory behavior between the two individuals. During late estrus the consorting musth male lost interest, and she was once again chased and mounted by younger males.

When younger males attempted to mount females during early and late estrus, long chases and loud bellowing by the female were involved, which attracted other males. The noise also attracted the attention of human observers, which further led casual observers to believe that the younger males were responsible for the breeding. Estrous females avoided these younger nonmusth males by walking away and, if caught and mounted, they would simply walk forward, causing the young male to fall off. Younger males had a further disadvantage; it seemed to take some years of practice before they learned to control their highly mobile penises.

The behavior of a female around a guarding musth male was notably different. The pair went into consortship, working together to ensure that the younger males did not get a chance to mate. The movements of males around an estrous female were like a beautifully choreographed dance in slow motion. If the female took a step, her consort immediately took a step after her and all of the younger males moved out of his way. If a young male approached the female, she would move toward the musth male, soliciting guarding behavior from him. The musth male and estrous female mated only a few times during the two days they spent together. Typically, the guarding male would initiate a mating by pushing the female forward with his head, and then, laying his trunk across her back, he would mount. The estrous female would stand still for her guarding musth male, without bellowing; in many cases, she was the one who actually initiated the mating. Once the mating had concluded she let out a long series of very loud, low-frequency rumbles. Based on the behavior we observed, we postulated that ovulation and conception occurred during midestrus.

Males attempted to locate, guard, and mate with as many estrous females as possible during their musth periods. They were able to guard only one female at a time and they spent much of their time searching for receptive females. On one occasion I observed Bad Bull guarding two females together. He was able to accomplish this feat because the two females were experienced matriarchs from a single bond group who worked together to ensure that he was successful.

During the wet season, females gathered into large aggregations,

increasing the probability that an estrous female would be found by a musth male. Estrous females also assisted the searching males by calling loudly, by producing urine with particular olfactory components, and by their conspicuous manner of walking.

During the spring of 1982 I went into an intensely manic state, working until 2:00 A.M. most mornings and returning to the office again at 6:00 A.M. In mid-July I was able to submit my dissertation, entitled *Musth and Male-Male Competition in the African Elephant*, having completed it in a record nine and a half months. I think that many people who knew me well looked upon my time at Cambridge as emblematic of one aspect of my personality: an ability to shut almost everything and everyone out and keep my attention focused on the goal or the object of the moment. It was a determination bordering on obsession, and Cynthia referred to it as "Joyce putting on her blinders." My own priorities and way of doing things have often been strikingly different from those around me, and I have been made very aware that I have, on occasion, irritated and even angered some of my friends. I think perhaps I have mellowed over the years, but the reaction of some people also has upset and hurt me. Part of me feels that it *is* my fault, that I should be more considerate and not so self-absorbed. But another side of me says, I am who I am and we each have our own lives to live. I believe, too, that true friends accept those they love for who they are, and I feel fortunate to count quite a number of those special people in my life.

In any event, in July I sold my Mini for $20 and departed for Amboseli, with plans to return for my oral examination in October. The original plan had been for Keith and Pili to carry out their new studies while I was writing up my Ph.D. thesis, which meant that they would be completing their research just as I would be about ready to start another project. My study of musth had given me insights into the sensory world of elephants, and I was keen to begin investigating their vocal communication. As it turned out, Pili and Keith were only just starting their fieldwork when I submitted my dissertation and, due to a new rule made by the Kenyan authorities, there could be no more than two people on any research project at any one time. Therefore, although I was awarded my Ph.D. later in the year, I was forced to delay my plans. Though I was prevented from embarking on a study of elephant communication, I stayed on in Amboseli and focused my energy in an entirely different direction.

CHAPTER ⇒14

A Maasai Named Meloimyiet

THROW YOUR HEART OUT IN FRONT OF YOU
AND RUN TO CATCH IT.
—Anonymous

IN AMBOSELI, I HAD a Maasai companion named Meloimyiet. I don't remember when I first met him; his was one of those beautiful, sculptured faces that had always been a part of Amboseli, and then, one day, several months before I departed for Cambridge, he became a part of my life. Meloimyiet had almond eyes, dark velvet skin, and the slow, sensuous movements of a dancer. He was of proud Maasai temperament, often disdainful of other people, and as vain as an ocher-painted warrior. One moment he could be warm and charming, his eyes sparkling with good humor, and the next he was withdrawn and sullen, his face as silent and as black as the night. With Meloimyiet, I experienced a world of sensuality and passion, discovering new meaning in everything I did. For me, being immersed in a new culture, in a new way of looking at life, was intoxicating and almost addictive, while it lasted.

Meloimyiet had four small parallel lines on either side of his forehead, incisions that had been made when he was a child suffering from headaches, to let out the bad blood. Just below his cheekbones he had two circular scars, brands that prevented eye disease and blindness. On either thigh were several large circles of shiny skin: Decorations, he said, a sign of beauty. He and his age-mates had made them when they became warriors. Rolling up a piece of cloth, he had smeared the end with sheep fat, held it over the fire, and pressed the sizzling circular end against his skin. Like many Maasai of his age and background, he had pierced and stretched his ear lobes at puberty, when he was still young enough to believe that he would spend his youth as a warrior decorated with colorful beads. Once he was older, his ears became a source of embarrassment, because they were a sign that he had not had a formal education.

But Meloimyiet had what so many of us have lost: an intimate knowledge of and feeling for the natural world he lived in. When we were together in the bush, he taught me how to read elephant signs in the dust, to learn which way they had gone, what they were doing, how many had past, and how long ago. When the storm clouds gathered over the dry, dusty plains, he taught me to cup my hand toward the rain-bearing clouds in a receiving gesture, for to point at them would chase them away. During the rains, when the grass grew long and green on the *Eremito* ridge and the small pink and white flowers covered the plains, he taught me which species of grass gave the cows the most cream and which plants small boys ate when they were out with the cattle all day. He taught me which trees could be used for the treatment of worms, which for gonorrhea, which for fever. He showed me how to make a toothbrush from the *mswagi* bush and how to make fire with two sticks.

Meloimyiet also taught me the names of all the wild animals and the many names for the elephant: *Oltome* and *Olkancaoi*, the big one; *Olenkaina*, the one of the hand; and *Esalishoi*, the wild and fierce matriarch. He told of Maasai rites of passage. As the Maasai advance through different stages in life, they are given new names in celebration of this passage. A boy child is given a name at birth, another when he has distinguished himself as a warrior. Siblings refer to one another by the gifts they have exchanged, such as *endouwo,* a heifer, *esupen,* a sheep. Women, too, earn different names at different stages of life. They are given a new name when they leave their own families to become members of another and later become known as the mother of their eldest child. Meloimyiet, too, was known by several names: Onina, the loved one, was a pet name used by his mother; Ndouwo, heifer, the name used by his elder brother; Ole (son of) Nkurupe, that used by acquaintances; and Daniel, his Christian name.

He told me stories from his childhood, of the many adventures he had had on the long days out herding cattle, goats, and sheep. Once, when he was seven, he had fallen asleep and awoken to find himself surrounded by elephants. Ever so slowly he had taken off his *shuka,* the cloth he tied over his shoulder, laid down his sticks, and crept away naked between the animals' pillarlike legs. One day he had been thrown high in the air by a python, and one night he was tracked through the bush and almost eaten by a lion. He told me, too, of a woman he had known who had given birth to a bird.

Meloimyiet taught me about the protocol and intrigue of love among the Maasai: that there are strict penalties for the uncircumcised boy (*olayioni*) who is caught making love to a circumcised girl (*esiankiki*), though they may be of the same age, while there are no such penalties for the *olayioni* who makes love to an even younger but also uncircumcised girl. It is a cause for great shame if, during the *eunoto* ceremony, when a warrior becomes an elder, a mother cannot shave the head of her son, for this means that she had sexual relations with the warriors of her son's generation. Though men may have several wives and many girlfriends, women have secret lovers outside of marriage. He told me stories of his own many lovers, of the many circumcised *siankikin* he had slept with when he was still an uncircumcised *olaiyoni.* He told me, too, how he and his age-mates had burned tortoiseshell in the girls' houses so that they could sneak in and make love to the drugged girls without waking them. He told me of the days when they were recovering from circumcision and their mothers brought naked girls to stand before them. And how, when they cried out in pain and asked for the girls to be taken away, their mothers had laughed aloud.

I went to the circumcision ceremony of several young Maasai girls, not more than twelve or thirteen years in age. While men and other young girls danced and sang outside, these girls sat bleeding in the darkness of their house, looked after by the woman who had just mutilated them with a razor blade. The girls each sat in a darkened corner on a cow-skin mattress leaning against the wall, their knees drawn up and their legs spread. They had come of age. I found myself wondering how women could continue to mutilate one another's bodies generation after generation. Young girls who had already tasted the pleasures of sex had had their entire clitoris and sometimes labia minora removed. The whole idea of female circumcision was appalling to me, but I tried to understand it in the context of the Maasai and their culture. I spoke to my close Maasai women friends, some of whom assured me that they still experienced orgasm, but I wondered how their experience could be anywhere near as pleasurable as it had been beforehand. Many years later, when I was visiting Cyn at Lake Jipe in Tsavo National Park, we listened to the endless screams of a six-year-old boy of the Akamba tribe who was being held down and forcibly circumcised in the pump house. The experience left me appalled by male circumcision, as well.

At night Meloimyiet told me Maasai stories and legends including a tale of how elephants had once been people. Many years ago, he told me, there was an *esiankiki* who was leaving her *enkang* to be married to a man who lived far away. Two men from her future husband's settlement were waiting to take her there. They would have to walk for several days to reach their destination; the two men walking ahead of her, while she followed solemnly and obediently behind. She was tall and beautiful, with young breasts that stood up above her beaded leather skirt. Layers of colorful glass beads encircled her neck, and strings of beads hung down to her feet. It was the start of a new life; she had to go, but she did not want to leave her family. Finally she bid them farewell and turned away. She walked past the thorn fence of the *enkang* and then paused to say good-bye to her mother just one last time. She turned her head, and because of that failing she became an elephant. To this day as a young Maasai girl leaves to be married she must not turn to make her farewell to her family a second time. This, the Maasai believe, was how the elephants began. It is why the elephants' breasts look like those of a young girl and why the Maasai honor dead elephants in the same way they honor dead humans, by stuffing leaves or long stems of grass into the orifices of their skulls, much like flowers on a grave.

My time with Meloimyiet also taught me of isolation and racism, of my own emotional limitations and my ability to overcome. Kenya still has some difficulty with interracial couples, and perhaps ours drew more attention because we were from such different backgrounds. In Nairobi, at parties, people simply seemed at a loss for words, as if I were a girl they knew who had "gone wrong." And in the city I was painfully aware of the heads that turned to stare as we walked down the street together.

My family and friends were not pleased by my new relationship. Parents always want the best for their children, and though my mother kept most of her feelings to herself, I knew that she was disappointed and worried by my newfound passion and understandably anxious about my future. My friends, too, showed their concern. Camp was a small, intimate place, and some of the others found it awkward sharing it with Meloimyiet, who was of such a different background and a difficult person by any standards.

When I departed for Cambridge in 1981, Meloimyiet remained in camp for a while. Eventually he returned to *Ol Tukai,* and although I was not told that there had been misunderstandings, I sensed from Cynthia's remarks that problems had arisen. Our separation was made more difficult because there was no telephone contact and, with Meloimyiet's limited education, we corresponded with difficulty and usually through a third party.

When I returned to Amboseli the following July, I expected Meloimyiet to be waiting for me. He was not. Cynthia told me then that he was no longer welcome to stay in camp and that if I wanted to see him, I would have to go to *Ol Tukai*. The statement was made tersely, and I knew that it did not invite discussion. By national park standards *Ol Tukai* was a slum. Small, one-room shacks served as quarters for the parks and lodge staff. Rubbish lay everywhere and was even caught in the thorn trees. Making the choice between staying in camp with my friends and spending time with Meloimyiet in *Ol Tukai* was difficult, and I was terribly hurt then by what seemed like banishment. I was also hurt that no one would tell me what was wrong. Not until many months had passed did I learn that there had been a disagreement between Meloimyiet and a new camp researcher. When asked about the incident, Meloimyiet was speechless and told me that he had just been ordered to leave and that no one had even offered to help transport his things. He had had to walk to *Ol Tukai* to search for a vehicle to assist him. I will never know what really happened, but I think that the friction had more to do with people's unease with our relationship than with any incident that may have taken place. In retrospect, I completely understand the discomfort they felt in his presence, though I am still saddened by the rift that exists between cultures.

Other friends showed their concern in quiet ways. Always loyal, Cyn never commented on the matter but joined Meloimyiet and me in our various walking safaris, Maasai ceremonies, and adventures, though I could sense her hope that it was a stage I would grow out of. Pili and Keith, too, were tolerant and supportive, including us at dinners and on outings and talking me through the difficult times.

If I found it hard to cope with the rejection of my friends and acquaintances, I found it even more difficult to come to terms with the complete acceptance of his family and friends, who were warm and welcoming, treating me as if I were one of their own. The situation was potentially complex, in that Meloimyiet was married with two small sons, Sabore and Sokoine. Despite the fact that I knew (and was constantly reminded) that Maasai men were free to take several wives and that, if their wives were discreet, they, too, were allowed, theoretically, to have several lovers, I couldn't believe that Meloimyiet's wife, Simaio, could possibly appreciate my appearance in her husband's life. But Simaio was always gracious and kind to me, welcoming me into her home and often coming to visit me in Amboseli or sending Sabore, whom I adored, to stay with me for a week or two at a time.

While I stayed with Simaio or one of Meloimyiet's sisters, I was able to experience the way a Maasai homestead works. Though most of the family no longer lived in the traditional Maasai houses, his mother, Sikoi, still did, and Simaio and I made the kilometer-long journey to her house every morning and evening to collect milk and to see how she was. Evening was when the cows and goats came back from a day of grazing, and I was fascinated to see how the Maasai "count" their animals. Rather than tallying the number of animals, they remember each individual and check to see whether it is there or not. This takes considerable time when there are over a hundred goats, and I still don't understand why they don't just count. I had two of my own goats with their herd. One I had rescued from being slaughtered in Amboseli, because for some reason I decided I liked it, and the other had been given to me by Meloimyiet's mother. I was always asked whether I could recognize my goats, one of which was rather nondescript, and when I made a mistake everyone thought it was terribly funny.

Other domestic activities included collecting drinking water from the river, where we also washed our clothes and bathed, as did everyone else in the area. We cooked over a simple charcoal stove, or *jiko*. When it came to eating, I discovered that men of the warrior or junior elder age-set and women cannot eat together. And a warrior cannot eat anything unless he is accompanied by another warrior, which encourages strong bonding between these young men. The men ate outside while the women ate indoors by the light of a small tin kerosene lamp known as a *koroboi*. I learned, too, that many people are expected to sleep in one bed. I can vividly recall one night when I slept head to toe on a narrow single bed with Simaio, Meloimyiet's younger sister Naneu, and a child of five. I was placed in the middle until I started to feel unbearably claustrophobic and requested to sleep on the edge. There I balanced, wide awake on my side, for the rest of the night. Another night I shared a traditional leather mattress bed with Meloimyiet's older sister, Pialo, her three-year-old son, and her newborn. Next to the bed were some chickens that scratched all night and some sheep that made the most dreadful sneezing sounds. Maasai babies don't wear diapers, and as leather mattresses are not absorbent, the urine ran down toward my side of the bed several times each night.

Though my friendship with Meloimyiet's family remained strong, my relationship with him began to deteriorate. He later claimed that it was my fault and that it had begun after I had visited the United States, when I had made it clear to him that our relationship did not have a future.

Meloimyiet worked in Amboseli as a tour driver–guide, following in the footsteps of his elder brother, who was an Amboseli ranger. Like so many of his age-mates, he was caught between two cultures, trying desperately to emulate the mode of dress and worldliness of his clients and friends and yet unable to shed his traditional Maasai heritage. I think, with time, Meloimyiet came to see me as a way of escaping from his relatively poor Maasai background. Though he spoke with derision of the other young Maasai who had abandoned their families and lost their way in the coastal resorts of Mombasa, he, too, was drawn like a magnet to the material side of my Western culture, and I began to worry that he was headed in the same direction. I had hoped that he would eventually be able to support his family and improve their standard of living, but instead he began to live for the city life, particularly the discos, and would come back from frequent trips to Nairobi dressed in the latest fashions, wearing strings of gold necklaces and eventually with sticky, shiny permed hair. What I did not realize then was that he also was involved with a growing number of women, and perhaps his entanglements added to the tension between us. As his forays grew longer and more frequent, I began to realize that I was the only reason he continued to return to Amboseli. I began to feel trapped, wanting to escape and yet not wanting to leave him for fear that he would then abandon his own family totally. I also felt that I must try to recoup some of the money I had given him. In the end all of my attempts to talk sense into him failed, and Meloimyiet took off for Mombasa and never returned.

I have always felt partly to blame for his abandoning his family, though rationally I know that although I exposed him to a different way of life, he was the one responsible for making his own choices. A different man would have chosen differently. I still keep in touch with his family, though I see them very infrequently. They admonish me for not coming to stay, and I think they feel that, like Meloimyiet, I have abandoned them.

In many ways I feel terribly sad about what happened and I wonder if, given the chance at the same stage in my life, but with hindsight, I would do it again. Despite the pain and the scars, the richness that the experience gave my life, the understanding that I still carry with me from our time together, the sense it has given me of being a part of a land and a people is too important to me to write off as simply a foolish love affair.

PART THREE

REFLECTIONS ON ELEPHANTS

CHAPTER ⇒ 15

Elephant ESP

IT IS SAID THAT ELEPHANTS TALK TO ONE ANOTHER,
MUMBLING WITH THEIR MOUTHS THE SPEECH OF MEN.
BUT NOT TO ALL IS THE SPEECH OF THE BEASTS AUDIBLE,
BUT ONLY THE MEN WHO TAME THEM HEAR IT.
—Oppian, *Cynegetica,* II

IN MAY 1984 I turned twenty-eight years old, and wondered whether it wasn't time to leave Amboseli, but with some misgivings I embarked on yet another elephant study. Since 1982 I had been eager to investigate the elephants' vocal repertoire, but I was still fascinated by the various vocal and olfactory signals associated with musth, and I was intrigued to know how males used these signals to locate or avoid other musth males and to assess one another's fighting ability. Though I was torn between the two pursuits, in the end I was lured back into further work on musth, but I intended to spend part of my time recording elephant vocalizations. Keith and Pili were winding up their research when I applied for funding to the Guggenheim Foundation for fieldwork, which I hoped would begin in July. I contacted Daniel Rubenstein at Princeton University to inquire whether he would be willing to have me as a postdoctoral student. A warm and energetic man, whom I had first met in Cambridge when he was a Fellow of King's College, Dan was full of ideas and enthusiasm. Like me, he was interested in fighting strategies, signals, and assessment, and agreed to supervise a further investigation of musth. I was fortunate; the funding from the Guggenheim Foundation came through and after spending several months at Princeton discussing my plans with Dan and other colleagues and purchasing the necessary equipment, I returned to Amboseli in late 1984.

Out watching elephants once again, I had a strong sense of coming home. Though I had spent very little time with the elephants over the previous two years, their postures, actions, and responses to other individuals and to situations were so familiar to me that I slipped back into my life with them as if I had never been away. If anything the time apart had sharpened my senses, and I began to notice behavior that I had previously missed and to ask further questions of behavior that I knew well.

For example, there were times in Amboseli when I would watch a group of elephants all lift their trunks high in the air, responding to some scent that I could not detect. And other times I would see a matriarch, taking up the end of the line, stop dead in her tracks and spread her vast ears, listening to some sound imperceptible to me. The members of her family, without looking back, froze in unison, waiting for her to indicate what they should do next. How, when they were facing away from her, did they know that she had stopped? Then, as she lifted her head and opened her mouth, her family appeared to listen to a message that only they could hear, urging them to run away in silence, their heads and tails high. A naturalist working in Tanzania by the name of A. F. Rees once called this uncanny sense of elephants ESP. In a sense, he was right for, at least when measured against those of humans, their perceptions are extrasensory.

A friend once said to me: Imagine that you presented drawings of an elephant head and a human head to someone from Mars. What could he tell you about the two species? That the one with the long nose and big ears must have an incredible sense of smell and hearing, and that the other, with the tiny nose but proportionately large eyes, was probably rather poor of smell but must have relatively better eyesight. How right he would be. Elephants are so very different from us that they live in effect in another sensory world, hearing things inaudible to us and smelling things we can't detect.

When I left camp each morning I habitually rolled my car windows down, no matter how cold it was, for with the window up it seemed to me to be more difficult to find elephants. As I drove through *Oltepesi, Ol Tukai Orok,* or *Kitirua,* I would get a certain sense about whether elephants were in the vicinity. If elephants were absent, the landscape had a stillness, an emptiness. If I was with another elephant person, I always voiced my thoughts: "It feels as if there are no elephants here today," and my companion would always agree. Once we acknowledged that we never did find any. If elephants were nearby there seemed to be a vibrancy in the air, a certain warmth. Now I began to ask myself whether it was possible that over the years our senses had been trained to pick up sounds and smells of which we were only subconsciously aware, or whether our intuition was in fact just our imagination. I wondered whether we were commenting on the presence or absence of elephants only after considerable searching and whether we would have failed the test if we had been blindfolded.

It was the long hours of watching musth males that gave me real insight into the sensory world of the elephants. A musth male might very carefully test some invisible spot on the ground and then, with his trunk sweeping back and forth just a centimeter above the earth, meander along the trail of another elephant who had passed hours, perhaps even a day, before. Or a musth male might raise his head, his ears spread, his body motionless, and then suddenly, in a very determined way, walk in the direction he had been listening. It was this kind of walk that we called *anajua pahali anapoenda,* "he knows where he's going—and we don't." The frustration, but also the excitement, of watching elephants was trying to second-guess their behavior. I made a game of it: Whom was he listening to? A musth male rumbling, or a female in estrus calling? Whom was he going to meet, where, and why? I tried to discern the subtle differences in the male's behavior, made my predictions, and then followed, sometimes for hours, to see whether I was right or wrong.

One of the vocal signals that musth males use is an almost inaudible sound, like water gurgling through a deep tunnel, which I call the musth rumble. When I first heard Oloitipitip rumble in 1980, I had mistaken it for the sound of his vigorous ear-flapping. Only after watching and listening closely did I realize that it was, in fact, a vocalization. As time passed, I discovered that each of Amboseli's musth males used this particular rumble as part of a threat display associated with an unusual sort of ear-flapping that I call an ear wave. It seemed strange that an aggressive call made by males of world's largest land mammal should be barely audible, and I began to think that the musth rumble was probably extremely low in frequency, and that perhaps I was only hearing part of the sound.

To understand this vocal signal better, I proposed to make audio recordings of the rumbles of particular males, play them back to higher- and lower-ranking musth and nonmusth males, and observe their responses. Playbacks can be a very powerful tool in the study of animal behavior, as they provide, in effect, a way of asking animals questions about their social world. I predicted that if I played a musth rumble to another musth male, he would approach the speaker in an aggressive manner, while if I played the same call to the same male when he was not in musth, he would avoid the sound.

I knew from experiments carried out by Rickye and Henry Heffner that elephants are capable of hearing sounds of very low frequency. It seemed logical, therefore, that they could also produce them. In order to carry out my experiments, I would need to be able to record and play back some very low frequencies, and the equipment would need to run off twelve volts, so that I could power the system from my car. While I was at Princeton I searched for the appropriate audio equipment, trudging up and down the streets of New York City to where the best of the rock groups shopped. As I walked from store to store describing what I required,

I began to realize that I had a tall order. Most of the available audio equipment did not perform well below about 20 Hertz, since humans cannot hear below 30 Hertz, except at high sound-pressure levels.

While I discussed my problem with colleagues at Princeton, someone suggested that I contact Katherine Payne, a dedicated biologist who had spent many years studying the vocalizations of whales and was intending to embark on a study of the vocal communication of Asian elephants. When I spoke to Katy over the telephone in September 1984 she had just returned from a visit to the Washington Park Zoo in Portland, Oregon, and she had an exciting hunch about elephant vocalizations.

I described what I was proposing to do and explained my problem in locating equipment. She countered my questions with one of her own: Did I think that African elephants were making sounds below the level of human hearing? Cynthia had long believed that elephants make inaudible sounds, and I recounted to Katy that we frequently noticed behavior suggesting that elephants were reacting to things we could not hear and that they sometimes opened their mouths wide as if vocalizing, though we heard no sound. I told her, too, of my barely audible, six-ton, aggressive musth males. Then I asked what was behind her question, and she recounted her tale. While she had been sitting watching the Asian elephants at the Washington Park Zoo, she had felt deep vibrations, although she was aware of no audible sound. She was sure that the elephants were producing the vibrations. With her long experience listening to the low-frequency calls of whales, she had guessed immediately what this might mean: The elephants were communicating with one another using infrasound, sound below the level of human hearing.

Being so familiar with elephants, I did not immediately grasp the significance of what she was saying. But Katy explained that infrasound had properties that audible sound did not. Very loud, low-frequency sound attenuates more slowly through the environment than higher-frequency sound, which means that it can travel greater distances before dropping below the level of background noise. If elephants were using infrasound at high sound-pressure levels, many of our interpretations of their social world would have to change. For example, while we defined the members of a group visually, the elephants probably used sound, and if they could communicate with one another over several kilometers, their understanding of group composition was probably quite different from ours.

Katy told me that she was heading back to the Washington Park Zoo with a Nagra tape recorder that could record infrasound and that she would let me know if she found anything interesting in the Asian elephants. I was terribly excited by her ideas and suggested that if she discovered that Asian elephants were indeed using infrasound, then she should plan to come and spend some time recording African elephants in Amboseli with me. In the meantime, I decided that before I spent too much money on equipment, I'd better wait until I heard Katy's results.

Since I had been planning to study elephant vocal communication, I had already made a preliminary list of the different calls that Cynthia, Pili, and I had identified. When I returned to Amboseli in October, I began to take more detailed notes on the vocalizations and their context. African elephants make many different types of calls, from the higher-frequency trumpets, cries, bellows, screams, and snorts to the very low-frequency rumbles. Within each class of calls, there are several different types; for example, in the trumpet class, there are play trumpets, trumpet blasts, and social trumpets. The rumbles are the most varied and complex class of elephant calls. Originally they were referred to as stomach rumbles, as some hunters had assumed that these sounds originated from the elephants' digestive tracts. At the start of my research I was aware of at least ten different rumbles, each given in a different context and each having a different meaning: for example, the musth rumble, the greeting rumble, the suckle protest, the lost call, the estrous rumble, the contact call and contact answer, the "let's go" rumble, the female chorus, and quite a number of other "mystery" rumbles.

In late November Katy telephoned to let me know that she had found that the Asian elephants *were* using infrasound to communicate with one another. She promised to come to Amboseli for several weeks in January 1985.

When Katy came I moved from the house I had shared with Meloimyiet at the *Bandas* back into camp and found that the mood there had been transformed. Masaku had retired, leaving Peter Ngande, a tall, placid man in his late twenties, in his place as our cook. Peter's character had a calming effect on all of us; the atmosphere became relaxed. Though camp was often lonely, Peter never complained, and we returned after a long day in the field to a smile and a helping hand. Peter was later joined by Vincent Wambua, a wiry young man with a tremendous sense of fun. Wambua was a complete clown, imitating just about anything from windshield wipers to various camp characters. His imitations of animals were superb, and he was always telling us to *kaa ngiri*, "stand like a warthog"; be alert, ready for any eventuality. He could do a hilarious imitation of the duetting Bou-bou shrikes, flapping his arms vigorously and wiggling his rear end in the air. Within no time he had coaxed the swallows to nest in the kitchen, and when their nests fell down he built a special support for them out of an old Blue Band margarine tin and provided each bird with a roosting perch. Wambua had an incredible knack for building gadgets, and I always felt that if he had been given the opportunity to go to school, he would have been a very successful engineer.

I shared the camp first with Marc Hauser, who was studying vervet monkey communication and cognition, and, after Marc's departure, with Sandy Andelman and her husband, Rudolf. Sandy, who had completed her doctorate on vervet monkeys, had returned to Amboseli to study female elephant relationships. In 1987 Sandy's research took her to Manyara National Park in Tanzania, and as Cynthia was living mostly in Nairobi writing her book, *Elephant Memories,* I spent the majority of my time in camp during the late 1980s alone.

Katy arrived with an impressive array of equipment, most notably the Nagra IV SJ recorder, which had a frightening number of buttons, dials, and knobs whose function I had to learn. The Nagra consumed twelve of Kenya's "Ever Ready" batteries in a distressingly short period of time, and we frequently found ourselves on the far end of the park with no power. Katy and I spent the early hours of the morning, before the wind came up, with the elephants. Recording elephant vocalizations requires extreme concentration. We both listened intently, noting each vocalization we heard while I also recorded, to the best of my ability, which elephant had vocalized, which call had been given, and why. Very low frequency sound is difficult for our human ears to localize, and to complicate the matter, several elephants often vocalize in unison. Fortunately for me, elephants usually flap their ears when they call, which assisted me in making my educated guesses. Katy had the more difficult task of watching the group's behavior and ensuring that the levels of the two tracks of the recorder were set properly. Because some elephant sounds are extremely loud and some very soft, the two tracks were set at different sensitivities, and Katy had to listen alternatively to each track to ensure that we were not getting too much wind noise. In one ear Katy listened through an earphone to the tape, and in the other she listened to the elephants. Her job was made even more confusing because she was recording at the slowest possible speed, which meant that the vocalizations she was hearing through the earphone came a second or so after she heard the same sound through her other ear.

Katy had designed some data sheets specifically for our task, and we compared our notes once we got back to camp. During one twenty-minute stretch of recording, Katy heard twenty-two rumbles, while with my trained ears or perhaps better low-frequency hearing, I had heard thirty-four. But when we played the tape back at ten times the recorded speed, bringing the elephants' sounds into audible range, we counted 102 calls: Clearly the elephants were making many sounds that we could not hear.

Once we analyzed the data and produced sonograms, we found that most of the elephant rumbles contain harmonics, frequencies that are multiples of the fundamental, or lowest, frequency. While the fundamental frequency was almost always inaudible, in many cases the upper harmonics were well within the audible range. But as many as two-thirds of the sounds we recorded contained *only* frequencies below the audible range. The question remained: Were these inaudible sounds made by elephants who had been standing near the recorder, or were most of the sounds we hadn't heard been made by elephants who were some distance away? In the first case this would mean that elephants were vocalizing more frequently than we had believed and that most of their sound was inaudible to humans, which would make our research extremely difficult. On the other hand, it was possible, and more likely, that as the loud vocalizations made by more distant elephants traveled through the environment, the higher frequencies had attenuated below background noise, leaving only the inaudible frequencies to reach Katy's recorder. In this scenario, the elephants we were with could hear their distant companions while we could not.

Katy's visit had been very fruitful, and she made plans to return to Amboseli in 1986. Meanwhile, I knew that in order to record *and* play back elephant sounds I was going to have to go back to the States and work with Katy's colleagues at the Laboratory of Ornithology at Cornell University to design an elephant speaker.

For the time being I concentrated on some of the other musth signals. I wanted to study urine dribbling in greater detail, and I decided that I needed an assistant. The people of *Ol Tukai* were always searching for work for their many relatives, and when the word went out that I needed help, Zippora, the housekeeper at Amboseli Lodge, came to offer her sister, Norah Njiraini.

Norah came to the interview wearing a brown skirt and shirt, her long, soft hair combed out in a halo around her head. She was a slender young woman, her skin color *maji-ya-kunde,* the water of cooked lentils, not very black and not very brown. She had soft eyes and a lovely smile that revealed a big space between her white teeth, a sign of beauty. Norah was shy, and, with Zippora answering all of my questions, I decided that the only way to find out whether she was the right person for the job was to take her to see how she reacted to the elephants.

The following morning we went out together to *Longinye,* where we found Alfred and Pablo in musth, both interested in Ulla's nine-year-old daughter Ute. In those days Alfred and I were still testing each other's nerves, and he had a habit of coming right up next to the car, towering over its hood, and rumbling at me. While I was almost sure that I could trust him not to do anything unpleasant, there was enough doubt in my mind that I felt a rush of adrenaline every time he stepped near. Being approached by Alfred would be a good way of assessing Norah's courage and resolve.

On that day Alfred went through the complete musth male routine. He paced back and forth, mud-splashed, ear-waved, musth-rumbled, and dragged his trunk over his forehead and temporal glands. We watched from a safe distance as I explained to Norah the different behaviors we were observing and described what would be expected of her in her new job.

All of a sudden, and for no apparent reason, Alfred came for us, heading toward Norah's side of the car. My policy with musth males was to hold my ground, except with Bad Bull. As he moved nearer I saw Norah's hand reach out to wind up her window. I told her not to worry and to just hold still. She very wisely chose to ignore me. Alfred reached his trunk out toward her and, as Norah steadily wound up her window, he pressed his trunk up against the glass, leaving a stream of mud and mucus oozing down the pane. He towered over the car, rumbling at us with his ears folded, and paused for a terrifyingly long time before he strode off after Pablo. My hands were shaking so badly that my field notes from that day are almost illegible. I tried not to let on to Norah that I had been frightened, and I could detect no sign of fear in her.

On the way home I asked whether she had been scared. "When?" she asked. "When Alfred tried to put his trunk through your window," I said. "No," she answered, "you weren't scared and I knew that you would know if it was dangerous." I decided then that I would hire Norah, that she had what it took to study musth males.

Two days later she asked, "Joyce, do you think that I can really do this job?" Her question sounded suspiciously like a response to the standard *Ol Tukai* criticism inspired by jealousy: If you couldn't succeed yourself, it helped if you could at least ensure that others failed. I asked her what the talk was in *Ol Tukai*. She told me that people were saying that studying elephants was definitely not a job for a woman and that she would certainly fail. I explained to Norah that the opposite was true, that most behavioral studies were performed by women. Success in her job required only an interest in the elephants, dedication, and hard work, and with those qualities she would succeed.

Over the next five years we spent eight hours in the car together each day, following elephants, talking about life, telling *Ol Tukai* stories, laughing and crying. Living in the bush gives one long spans of time in which to contemplate life, and although we came from very different backgrounds, we became close friends. We had very different personalities, and I found in Norah a quiet strength on which I could depend.

Norah and I followed musth males day after day, collecting samples of urine-soaked soil from high-ranking musth males and presenting them, in Petri dishes, to lower-ranking musth males. I predicted that males were able to assess the owner of the urine. We filmed each male's response, looking for changes in behavior, direction, or movement of the trunk as clues to what was going on in its brain. Often the male stopped dead in his tracks, reversed himself for several steps, searched the ground with his trunk until he found the urine-soaked soil, and carefully tested it for several minutes. Then the male would hold still and listen and then, raising his trunk high in the air, search for the higher-ranking male.

Part of Norah's job during the three-hour focal samples we carried out was to write down, every minute on the minute, the musth male's activity and the amount of urine he was dribbling on a scale of 0 to 10. We then simulated the different rates of dribbling at a tap in *Ol Tukai* and calculated that the older musth males were losing an average of 240 liters of urine every twenty-four hours! Since males were clearly dribbling urine in order to leave messages for other males, I expected that they would adjust the amount of urine they passed depending on the activity they were engaged in. And indeed, this proved to be the case. While resting or standing and feeding, the amount of urine a musth male dribbled decreased, but when the male was walking or interacting, that amount increased dramatically.

Katy returned in early 1986 for another month of fieldwork, bringing with her a Nagra tape recorder for me, funded by the Guggenheim Foundation, which was still supporting my study.

In late 1986, Katy, her colleague Bill Langbauer, and I worked together at Cornell on the data we had gathered. We found that all of the rumbles made by elephants contain fundamental frequencies between 14 and 35 Hertz, and some had sound pressure levels as high as 102 decibels (measured at five meters from the source). To give some idea of how "loud" some of these sounds are, the highest sound pressure levels produced by elephants fall into the categories of "intolerable" (for example, a construction site is 110 dB and a shout measured at 1.5 meters is 100 dB) and "very noisy"(e.g., a heavy truck measured at 15 meters away is 90 dB and an urban street is 80 dB)—as defined by Thomas Rossing in his book, *The Science of Sound*. Calls with the highest sound pressure levels included the postcopulatory call, the mating pandemonium, the female chorus, the greeting rumble, and the contact call and answer, each with sound pressure levels falling in the range of "intolerable" and "very noisy." The long-term records collected in Amboseli on behavior and on the contexts of specific calls suggest that the elephants are making use of infrasound at high sound pressure levels in the spatial coordination of groups and as they search for mates.

During 1986 and 1987, Norah and I worked constantly, sitting silently in the car collecting elephant vocalizations on tape. It was about that time that Norah's sister-in-law, Soila Sayialel, a Maasai from nearby *Loitokitok*, joined the Elephant Project. Soila had a beautiful smile and eyes that danced with laughter and mischief. She also had better low-frequency hearing that either Norah or I, and in her the project gained another dedicated elephant researcher.

The repertoire of sounds grew to over thirty different calls, and still we knew we were only just beginning to identify them. Nevertheless, I put together a list of the calls we had collectively defined:

Sexual excitement
  Estrous rumble
  Female chorus
  Genital testing

Social excitement
  Greeting rumble
  "Social" rumble
  Roar
  Mating pandemonium
  Play trumpet
  Social trumpet

Group dynamics and coordination
  Attack rumble
  Let's go rumble
  Contact call
  Contact answer
  Coalition rumble
  Discussion rumble

Distress
  Lost call
  Suckle rumble
  Suckle cry
  Distress call (SOS)
  Reassurance rumble
  Calf response
  Suckle distress scream

Fear, surprise of strangeness
  Trumpet blast
  Snort
  Infrasonic alarm?

Social fear
  Scream
  Bellow
  Groan

Dominance
  Female-female
  Musth rumble
  Male-male

One interesting early observation that we made was that females use significantly more vocalizations than do males. Of the twenty-six documented vocalizations made by adult elephants, nineteen are made only by females, three are made by adults of both sexes, and only four are made exclusively by males. (An additional six calls are made only by subadults.) Of the twenty-two calls exclusive to females, nine are calls typically given in chorus with other family members, while thirteen are usually made by an elephant calling on its own. As far as we know, male elephants do not call in chorus. While females use many different vocalizations in active communication between and within family groups, males vocalize infrequently, apparently relying on listening to locate groups of females. By comparing the very significant differences in vocal repertoire of males and females, I gained a deeper understanding of the social worlds of elephants. The survival of females and their offspring depends on the cohesion and coordination of the family unit and the bond group and on their ability to compete with other groups for scarce resources. Most vocalizations made by adult females therefore are related to such family and bond group dynamics, while the few calls made only by males are related to male-male dominance interactions or reproduction.

I returned to Princeton in the autumn of 1987 for several months to write up a number of scientific papers and to commission Tom Danley of Intersonics to build an elephant speaker. It arrived at the Laboratory of Ornithology, where Katy was based, one day in late November: a two-hundred-pound elephant-gray box with two eighteen-inch motor-driven woofers and four eighteen-inch passive radiators (which helped the radiation of sound). Dave Wickstrom, a technological whiz and sound junkie from the Lab of Ornithology who worked closely with Katy's team, had advised Intersonics on the specification of the speaker and offered to test it for me. Over and over he played first 20 Hertz, then 15 Hertz, and finally 10 Hertz tones through the speaker at higher and higher sound-pressure levels, until even when I stood outside the house, under the stars of a clear November night, I could feel waves of nausea as the sound pulsed through my body. Late that night, on a high from the sounds and the success, we lay on the floor of his sitting room, feeling the beat of Mick Jagger's voice pulsing through our bodies in a way no one else has ever felt it before.

By the summer of 1988, I was in Amboseli, ready to begin play-backs. I was fortunate that Robert Seyfarth and Dorothy Cheney were there for fieldwork. Using their years of playback experience with vervet monkeys, they assisted me with the protocol for my experiments and worked with me for the first few days. It was not a simple task.

The audio equipment, including the massive speaker, was in No-rah and Soila's car, which was hidden from the subject elephant behind a bush. I was in a second car with a video camera, in a position where I could see both the elephant and the speaker. The elephant had to be a hundred meters away from the speaker and moving in such a way that approaching or avoiding it would re-quire a change in his direction. The audio equipment took four minutes to prepare, and Norah and Soila signaled to me when they were ready. I then signaled to them once I had started filming. After waiting sixty seconds, they played the recorded call through the speaker. I filmed for at least one minute after the playback or until the male had stopped responding.

In some ways I found these experiments terribly frustrating. I often felt removed from the animals, separated by a mass of elec-tronic equipment that all too often failed us. And frequently we had to set the equipment up several times because, when we finally were ready, the subject had moved into the wrong position. Every time the car was moved, each of the four passive radiators had to be secured with a bolt screwed through a protective wire mesh. But when we got our timing right and there were no loose con-nections, the results of the playbacks were very exciting. When we played a musth rumble to a nonmusth male, whether the subject was larger or smaller than the male whose call we were using, he stopped still and listened with his head raised, chin up in an appre-hensive posture, and then turned and moved away rapidly. When we played the same call to another musth male, the subject stopped and listened with his head high and chin tucked under, in an ag-gressive posture, and then turned and started walking toward the speaker with his ears folded. Musth males often stopped to musth-rumble themselves and then listened, presumably for a response.

Alternatively they searched for the nonexistent musth male by snaking the ground with the tip of their trunks and holding their trunks high in the air like a snorkel. Their ability to localize sound was so accurate that even from a hundred meters away the musth male would eventually come and stand next to Norah and Soila's hidden car as if to say, Well, this *is* where the sound came from, *where* is the male?

When we played the musth-rumble to a group of females, typically they would become excited, first listening and then answering with what we call a female chorus. And playbacks using the estrous call of a female caused males to turn on their heels and move rapidly in toward the speaker.

The experiments clarified the function of some of the reproductive calls I had been listening to for so long. Katy, her colleagues, and I had predicted, for example, that the estrous call, with a sound-pressure level of 102 decibels at five meters from the source, could be heard by other elephants perhaps five and maybe even ten kilometers away. The results of my experiments indicated that females might use this loud, low-frequency vocalization to attract males in from, let us say, a five-kilometer radius. Once gathered, male–male competition and her own selective tactics would ensure that she mated with the highest-ranking, oldest musth male, the male most likely to ensure that the male offspring produced would live to be large, healthy, and old, thus increasing the genetic material she passed on to future generations of Amboseli elephants. Working in Namibia, Katy and her research team did in fact test whether the estrous call could bring males in from some distance away. While half of the team observed a group of males at a water hole, the rest played the stimulus from a speaker positioned two kilometers away. After the call was broadcast, the males at the water hole left immediately and showed up at the speaker about half an hour later.

As my experiments progressed I continued to make audio recordings of other elephant vocalizations, and I became fascinated by some of the social calls used by related females. The level of communication was much more than just a series of instinctive or emotive sounds; instead the elephants were using many calls to elicit complex responses from family members and appeared, in some cases, to be carrying on simple discussions. The more I learned about elephant communication, the more I realized how very little I really understood, and I became intensely frustrated at my inability to decipher what the elephants were saying to one another.

CHAPTER ⇒16

Trunks, Tusks, and Tool Use

I HAVE ALWAYS BEEN intrigued by the number of different uses to which elephants put their trunks and wondered why such a useful appendage has not evolved in more than just the order of elephants, the Proboscidea. With so large a body, long legs, and so short a neck, elephants obviously had to have some efficient way of getting food and water into their mouths. The evolutionary answer was a trunk, or proboscis. The trunk of an African elephant is a highly sensitive organ equipped with an estimated 150,000 muscles and two small fingerlike tips at the end. It is at once a terrifically strong and yet highly tactile and sensitive appendage and is perhaps more versatile than a human hand. Elephants use their trunks to eat and to drink, to mud-splash and to dust, to comfort and to reassure, to fight and to play, to smell and to communicate.

Using its trunk, an elephant can push over an acacia tree or pick up a crumb a quarter of a centimeter in diameter. By sucking water into their trunks, elephants can then pour up to twelve liters of water into their mouths at a time. By flicking the tip of their trunks gently back and forth in a pool of water, elephants clean dirt and floating vegetation away. Mud-splashing techniques vary depending on the viscosity of the mud, but most often elephants scoop up mud in their trunks and, by aiming carefully, are able to cover almost their entire bodies. To dust, elephants use a slightly different technique: After scooping up some dust, they toss it toward a particular part of their body, blowing through their trunk as they toss so that the dust is spread evenly. They use their trunks, often combined with a trunkful of grass, to rub itchy eyes or ears. And if the inside of the tip of their trunk has an itch, they place it carefully over the tip of a tusk and twist it back and forth. Elephants use their trunks to reach back and touch a suckling baby, to calm one that is frightened, or to pull one out of harm's way. They use their trunks and tusks to try to lift a dead or dying elephant, and their trunks to examine the bones of the dead. An elephant greets a nonrelative by placing its trunk in the other's mouth. Elephants use their trunks to smell danger, to detect an estrous female from several kilometers away, to track a musth male, and to recognize the urine or temporal gland secretions of members of their own family. By holding their trunks in different positions, they communicate with other elephants, and by blowing through their trunks they can produce a variety of different trumpets and snorts.

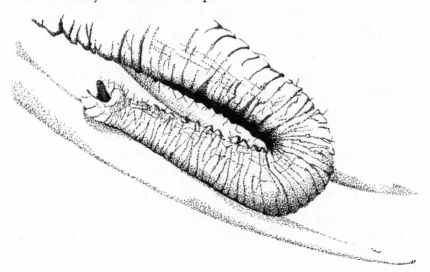

In studying the behavior of elephants, I have found watching the tip of the trunk to be highly informative. The tip of an elephant's trunk is almost never stationary, moving in whatever direction the elephant finds interesting. An elephant's attention usually is stimulated by what other elephants are doing, and by observing the trunk tip I often have been alerted to subtle behavior that is taking place in the group that I might otherwise have missed.

Many people have the impression that elephants stuff food into their mouths all day long, and they do spend a good sixteen hours a day feeding. They eat almost anything and everything that can be classified as vegetation: lianas, herbs, grass, reeds, leaves, thorns, bark, branches, or fruit. Elephants are herbivores, although we once witnessed Teddy eat an entire frozen chicken, and in Amboseli anything from rubber gloves, condoms, ladies' underwear, bottle caps, small bottles, batteries, and pieces of glass can be found in the dung of certain elephants.

Each species of plant is eaten in a slightly different way, and elephants use a variety of complex trunk, mouth, tusk, and foot movements to get a trunkful of vegetation into their mouths. For example, they grab papyrus and other swamp grasses and reeds with the trunk, pull them up complete with roots and mud, and place them in the mouth so that the muddy bit hangs just outside. This section is then bitten off and falls back into the water, while the remainder is gradually fed into the mouth with the trunk. Elephants grab *Cynodon dactylon,* the lush swamp edge grass, by the trunkful and stuff it into their mouth to be chewed and swallowed as another trunkful is gathered. Some of the short grasses on the plains are processed through coordinated movements of trunk and foot. Elephants kick at a stubborn clump of grass until the roots and stolons are loosened and then use the trunk to hold it firmly while a final kick removes it. Elephants don't like to eat too much dirt, and carefully clean a clump of grass by holding it in the tip of the trunk and scratching it repeatedly down the trunk's rough underside. Elephants may begin kicking at the next clump of grass as they clean the previous one. The longer grass in the *Acacia tortilis* woodlands is cleaned by holding it firmly in the trunk and shaking it against part of the elephant's body, usually between the forelegs.

Feeding on acacias is a quite complex maneuver, requiring co-ordinated movements of the tusks, trunk, mouth, and sometimes feet. The debarking of acacias and other trees involves first searching for a suitable spot with the trunk and tusks, and then prying a piece of bark off with a tusk and helping to remove it with the trunk. The softer-growing ends of acacia branches are simply encircled with the trunk and stripped bare of leaves and thorns for a length of eighteen inches or more. For longer, more mature growth, the branch is steadied with the trunk and bent and then broken over a tusk, a practice which over the years often leaves a deep groove in a favored tusk. The elephant then slowly feeds the branch into its mouth, so that each five-to-six-centimeter-long thorn is carefully bent over as it enters its mouth.

An elephant's trunk and tusks are its most useful tools, and many an elephant in Kenya has learned that tusks do not conduct electricity and can be used to break electric fence wires. But elephants also use tools that they find in their environment. On many occasions I have watched an elephant pick up a stick in its trunk and use it to remove a tick from between its forelegs. I also have seen elephants pick up a palm frond or similar piece of vegetation and use it as a fly swatter to reach a part of the body that the trunk cannot. Elephants have picked up objects in their environments and thrown them directly at me, undertrunk, with surprising, sometimes painful, accuracy. These projectiles have included large stones, sticks, a Kodak film box, my own sandal, and a wildebeest bone. In some cases these tosses have been a result of annoyance, in others, part of a game that the elephant or I has initiated. Elephants throw things at each other in the same circumstances: during escalated fights and during play. Elephants have been known to intentionally throw or drop large rocks and logs on the live wires of electric fences, either breaking the wire or loosening it such that it makes contact with the earth wire, thus shorting out the fence.

Young elephants must learn each of these different techniques, from drinking to feeding to manipulating objects. Babies often try to drink water with their trunks, but if they cannot do so satisfactorily, they resort to getting down on their knees and drinking the water directly. Mastering mud-splashing and dusting also takes practice, and it is several years before young elephants are able to do either with any proficiency. The same is true of feeding on the different species of grass, and I often have watched a frustrated baby try to use its trunk and feet to dislodge a clump of grass until it finally got down on its knees and bit off a mouthful. Babies frequently remove the grass from their mouths with their trunks and play with it for a bit, sometimes dropping it, sometimes throwing it, sometimes sticking it in an ear, and then occasionally returning it to their mouth and chewing it. Babies spend much of their early feeding eating the bits and pieces of vegetation that drop from their mother's mouth.

The amount of practice a young elephant needs to master the many uses of its extraordinary trunk is merely one dimension of the long period of learning elephants experience. One comparative measure of intelligence is the size of the brain at birth relative to its full adult size, which is considered to be an indication of the degree of learning a species undergoes during childhood. Among the majority of mammals this value is close to 90 percent. In humans, the brain at birth is a mere 28 percent of its full adult size, which, it has been argued, partly reflects the mechanical constraints of birth, but also indicates the long period of learning and social development that we undergo. Chimpanzees, our closest relatives, are born with 54 percent of their adult brain size. Elephants, too, are strikingly different from the majority of mammals with their brains at birth being only 35 percent of their full adult size, which in part must reflect the long period of dependency (about ten years) and learning that young elephants go through. The brain of a full grown adult elephant weighs between four and six kilograms, and the cerebrum and cerebellum are highly convoluted. The temporal lobes of the cerebrum, which in humans function as the memory storage area, are very large, bulging out from the sides of the brain.

Intelligence is very difficult to measure, especially in a species whose senses are so different from our own. Certainly elephants are intelligent by nonhuman standards, but watching their apparent concerns and interests, I can't help but wonder whether they experience emotions similar to our own and rudimentary conscious thought.

CHAPTER ➤17

Elephant Thinking

THEY MOVE FINISHED AND COMPLETE, GIFTED WITH EXTENSION
OF THE SENSES WE HAVE LOST OR NEVER ATTAINED,
LIVING BY VOICES WE SHALL NEVER HEAR.
—Henry Beston, *The Outermost House,* 1928

THE DEVELOPMENT OF CONSCIOUSNESS, language, compassion, and a sense of self are seen as true signs of humanity. The roots of these attributes, which evolved over several millions of years, can be found in our ape relatives, the chimpanzees. But what of other nonhuman species? Do some of the vocalizations used by elephants, for example, suggest that they are communicating with more than just signals, that they have a rudimentary form of language with syntax? Does some of their behavior indicate they have a sense of humor, a sense of death, a sense of self? And if elephants have a sense of self, how do they see themselves in relation to the rest of the natural world? These are difficult questions for which we do not yet have answers. But because elephants are complex social mammals, and because they often appear to act in ways that imply that they sometimes think and feel consciously and are self-aware, it is worthwhile examining the evidence.

One window into the way in which elephants think and feel is to assess what we know about their vocal communication. Our studies of elephant communication are still at a very early stage, and some of our interpretations of elephant sounds may well prove to be wrong. Only additional time watching elephants and recording, analyzing, and playing back elephant calls will correct our mistakes and allow us to clarify the many questions that remain.

One of the more fascinating vocalizations that elephants use is a low-frequency sound that we have called the "Let's go" rumble. When the matriarch of a family decides to stop feeding, for example, and move to the swamp, she usually flaps and slides her ears loudly against her body in what Cynthia and I have called a flap slide and then walks off. Other family members just have to pay attention and follow if they want to remain with the group. But when another adult female or younger family member wants to go somewhere else, she typically stands on the edge of the group, facing in the direction that she wants to go, often with one leg lifted and swinging back and forth, as if intending to take a step, and gives a soft, relatively unmodulated low-frequency rumble. If there is no reaction on the part of her family, she will wait several minutes and repeat this rumble over and over until the rest of the family either follows her lead or she goes off on her own. I think it would be fair to say that the Let's go rumble expresses an elephant's thoughts to the others: "I want to go in this direction, let's go together." Would it be accurate to say, then, that the Let's go rumble is a word? Sue Savage-Rumbaugh, who has studied the question of language in chimpanzees, has written that to qualify as a word, a communicative signal must have the following four attributes: (1) it must be an arbitrary symbol that stands for some object, activity, or relationship; (2) it must contain stored knowledge; (3) it must be used intentionally to convey this knowledge; and (4) recipients must be able to decode and respond appropriately to the symbols. I believe that the Let's go rumble meets all four criteria.

A related vocalization, or series of vocalizations, is one that I have called discussion rumbles. Elephants use those during times when they appear to be having a discussion, usually a disagreement, about the plan for the day's activities. A typical situation would be when some elephants in the family appear to want to pursue one goal, such as go to the swamp (and repeatedly move off and return), while the others seem to prefer to do something else (and move off in the other direction or stand facing in the opposite direction). The apparently debating rumbles (and there may prove to be several different types) go back and forth between the individuals, sometimes for twenty minutes or more, very much in the cadences of a human conversation, but in slow motion. Are the elephants actually conveying anything in a manner akin to that of using strings of words with syntax? Is one elephant actually sending a message as sophisticated as "I want to go to the swamp because I am hungry and thirsty," while another responds, "I don't want to go because I have a small baby that I cannot take into the swamp," for example? Although in some cases elephants can become quite agitated by the lack of accord in their group, I have also watched them use a similar but slightly different sounding series of rumbles in a nonconflict situation, to apparently "chatter away" to another individual. Could it be that the so-called discussion rumbles contain subtle differences in syntax and that by using such syntax, the elephants are able to convey a variety of different thoughts and feelings to one another?

The attack rumble, which I have heard on only a very few occasions, is another vocalization that, like the "Let's go" rumble, appears to function like a word. When a group of elephants feels threatened, it may do one of three things: turn and flee, bunch up in a circular defensive posture, or cluster in a pyramidal shape, with the largest females in the center front, and attack *en masse*. Apparently the matriarch usually takes the lead in deciding how to react. We have long assumed that when elephants flee, they are sometimes responding to an infrasonic alarm call given by a group member, but we have yet to record any sound that fits this description. In an offensive situation, however, elephants do exchange a very loud series of rumbles, quite unlike any others, which I have heard on several occasions when I have been under attack by a group. A large adult female, possibly the matriarch, initiates the rumbling, to which others respond by returning the rumble and clustering together, heads high and often touching, into a pyramidal

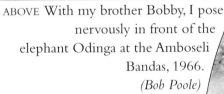

ABOVE With my brother Bobby, I pose nervously in front of the elephant Odinga at the Amboseli Bandas, 1966.
(Bob Poole)

RIGHT In my Hillview Infant School uniform, I proudly display a "nest work" pin at home in Malawi, 1964.

Bobby and I marvel at a pile of elephant dung while on safari in Kenya, 1963.

A typical scene: lunch on the hood of the Land Rover, Amboseli, Kenya, 1963. From left to right: my father, Bob; Bobby; my sister, Ginny; and I.
(Lee Poole)

My primary school class was made up of children from ten different countries. I am in the front row, third from the left. Kenya, 1967.

I share a cup of tea with friends at a cottage on the
Zomba Plateau, Malawi, 1963.

Standing outside my first
tent in Amboseli,
February 1976.

Cynthia Moss takes a
coffee break while
observing
elephants from a vantage
point on top of her old
Land Rover,
Amboseli, 1978.
(Joyce Poole)

The view from
camp looking
south to
Kilimanjaro.
(Joyce Poole)

The Elephant Camp,
January 1980. From left to
right: the shower, the
kitchen, Masaku's tent, the
dining room tent, my tent
(with roof under con-
struction). *(Joyce Poole)*

Outside my tent is the
"bathroom": water is available in a plastic container,
and soap is on the stool.
Toothbrush, toothpaste,
and shampoo hang in a basket from the makuti roof.

ABOVE After attempting to lift her stillborn baby for hours on end, Tonie stands near its small body in silence. It was watching Tonie's vigil over her dead newborn that I first got the very strong impression that elephants grieve. The expression on her face, her eyes, her mouth, the way she carried her ears, her head, and her body; every part of her spelled grief. Amboseli, 1980. *(Joyce Poole)*

OPPOSITE The power of a fight between two musth males is awe inspiring. Here Amboseli's highest ranking male, Iain, approaching downwind, catches Dionysius unaware. A dramatic battle ensues during which Iain lifts Dionysius off his feet. Dionysius wisely turns and flees, and Iain chases him across the plains for almost an hour.
(Joyce Poole)

BELOW As an elephant, with the AA family in the background, in camp, 1982.
(Keith Lindsay)

ABOVE A view from the Elephant Camp across the South Clearing: Cynthia observes the elephants while they feed and dust around her Land Rover. Snowcapped Kilimanjaro towers above. *(Joyce Poole)*

LEFT A large group of males rests together in Kitirua, Amboseli's western Bull Area. Three of my study animals stand in the front row on the far left: from right to left they are Ernie, Sleepy, and Andrew. *(Joyce Poole)*

Polly's death 1981. From top to bottom: Polly falls to her knees as temporal gland secretion streams down the sides of her face; she collapses and dies; two young males come and try to lift her, Kasaine pulling her tusks in an attempt to raise her head.

Having failed, Kasaine tries to rouse Polly by mounting her repeatedly.

After the rangers remove her tusks, Kasaine returns with two companions to pay his respects the following day. *(Joyce Poole)*

Masaku Sila, the temperamental camp cook, sits outside his tent in one of his better moods, displaying a chicken ready for the pot. *(Joyce Poole)*

Meloimyiet and I "dress up" in my collection of beads and "shukas," in camp, 1981. *(Joyce Poole)*

My field assistant, Norah Njiraini, Amboseli, 1985. *(Joyce Poole)*

Meloimyiet poses as a warrior for the record, in camp, 1981. *(Joyce Poole)*

Listening to music and looking at old elephant photograph albums outside Cynthia's tent, camp, 1986. From left to right: Cynthia Jensen, I, Conrad Hirsh, and Cynthia Moss. *(Joyce Poole)*

TOP LEFT My playback equipment in camp: a 200 lb. elephant speaker, a Nagra IVSJ tape recorder, earphones, and a 12 volt amplifier. Some elephant identification cards lie on the ground next to the "Female Box." Celeste and her family feed nearby, and Cynthia's tent stands in the background. *(Joyce Poole)*

TOP RIGHT Tania, Tallulah, and Tuskless's families display an intense greeting ceremony. *(Joyce Poole)*

MIDDLE LEFT I "rumble" to Joshua while elephant Joyce looks on, Amboseli, 1989. *(Bill Thompson)*

BOTTOM LEFT Young elephants play a game of King of the Castle while the older elephants take a morning siesta. *(Joyce Poole)*

ABOVE I watch the elephants, while the elephant Joyce watches me, Amboseli, 1989. *(Bill Thompson)*

ABOVE There is something eerie and deeply moving about the reaction of a group of elephants to the death of one of their own. Here elephants examine and carry the bones of another elephant. *(Oria Douglas-Hamilton)*

Joshua strolls past the author's tent examining the bones of another elephant. The dining room tent is in the background. *(Joyce Poole)*

The face of the ivory trade: a young male found by researchers in Amboseli, 1981. *(Joyce Poole)*

Oria and Iain Douglas-Hamilton
ready for take off. (*Gary Deane*)

Early days at Kenya Wildlife
Service: Richard Leakey visits with
Warden Kamanja (far left) and staff
at Mount Elgon National Park,
where poachers reduced the
elephant population to some 250
individuals. During a 1991 survey
of the Elgon elephants, the KWS
elephant team got caught in cross
fire between poachers and rangers.
(*Joyce Poole*)

ABOVE Richard Leakey
makes an impassioned plea
for the elephants, the Ivory
Burning, July 1989.
(*Oria Douglas-Hamilton*)

Barbara Tyack takes in the view
over the Rift Valley, Soit Naibor,
1992. (*Joyce Poole*)

ABOVE On the job and pregnant, I take measurements of an immobilized male elephant whose tusks have been sawed off in an attempt to deter him from breaking through electric fences and crop raiding. KWS veterinarian, Richard Kock, and two rangers work together, while a spotter plane circles overhead, Laikipia, 1992.

Holding the Amboseli Elephant Project identification cards, I explain how to identify elephants to Maasai elders and Community Game Scouts in Amboseli, 1994.

An interaction with a musth male. Beach Ball and I, Amboseli, 1994.

A KWS field trip: I hold four-month-old Selengei, with Tilly in the background, in camp, Amboseli, September 1993 (*Paula Kahumbu*)

Selengei, a little Maasai girl. Her necklace is a gift from Meloimyiet's family. Soit Naibor, 1994. (*Joyce Poole*) formation. The rumbling reaches a crescendo, until suddenly the entire group, including tiny babies, attacks. It is as if the elephants are exhorting one another, "Let's attack! Are you ready?" and then, perhaps, at some subtle change in the calls, "Attack now!"

Somewhat similar to the attack rumble is another call that I refer to as the coalition rumble, though we may find that these two vocalizations differ only in intensity. Related females use the coalition rumble to chase off annoying young males, an unrelated female, or another family group. Because a female on her own cannot drive away more than one elephant or any larger than herself, she elicits the assistance of another female or females in her family. Prior to the attack the elephants rumble, with their heads high and together, and also may clank their tusks against one another.

A baby or juvenile will use a distinct call when it is lost. The search for its family starts with a low-frequency rumble, quite different from any other rumble, and as the lost elephant becomes more and more frantic, its call becomes louder and louder, until after a time it ends with a scream. I have often lain in bed at night listening to the calls of a frightened calf, followed by the rumble of an adult, and finally by the sounds of the calf running off in that direction rumbling, trumpeting, and screaming.

Juveniles and adult females use two vocalizations that may develop from the lost call: the contact call and the contact answer. Females who are separated from their families use these two calls at varying sound-pressure levels to relocate their relatives or just to keep in contact with animals that are nearby but may be out of sight. Calling elephants are usually, but not always, answered by members of their family. Preliminary results from recent playbacks of recordings by Karen McComb, who is studying elephant communication in Amboseli, have suggested that the elephants who do answer tend to be more closely bonded to the caller than those who do not answer. My own observations suggest that the contact call may be more complex than that. I once observed an adult female call several times in the space of an hour and be answered first by another adult female in the group, then by her eldest daughter, and later by her youngest calf. I have watched a female separated from her family by a distance of two kilometers be answered by various different members of her family at different times but also by a nonfamily, nonbond group member. To the human ear, one contact call sounds much like another, except for variations in intensity, but with time we may very well find that there are other subtle differences between contact calls, and that a female may be able to call specific individuals. I am not suggesting that elephants call each other by name, but I do think that there may be subtly different sounds for "my mother," "my sister," "my eldest daughter," and so on.

As a behavioral ecologist, I have been trained to view nonhuman animals as behaving in ways that don't necessarily involve any conscious thinking and that their decisions have been simply genetically programmed through the course of natural or sexual selection. But in the course of watching elephants, I have always had a sense that they often do think about what they are doing, the choices they have, and the decisions that they are making. For example, when a young musth male is threatened by a high-ranking musth male, his usual response is to drop out of musth immediately. He lowers his head, and urine dribbling can cease in a matter of seconds. Many biologists would explain this phenomenon simply by arguing that males who behave in manner X live to produce more surviving offspring than males who behave in manner Y, and thus the trait for behaving in manner X is passed on to future generations. Thus, male elephants today automatically behave the way they do because they have been programmed through the successful behavior of their ancestors to do so. Although I rely on such explanations myself, as I have gotten to know elephants better I have been more and more convinced that they do think, sometimes consciously, about the particular situations in which they find themselves. In the case of the young musth male, I believe that he may actually consider his options: to keep dribbling, stand with head high, and be attacked, or to cease dribbling, stand with head low, and be tolerated. In other words, the male may in fact have some *conscious* control over the amount of urine he dribbles. With dominance rank between males changing on a daily basis, a male needs to be able to adjust his behavior accordingly. From past experience he knows the characteristics of his rival's body size, fighting ability, and how that rival normally ranks relative to him, but if his rival is in musth he also needs to assess whether he is in full musth and what sort of condition he is in. All of this information must be assimilated on a daily basis and gauged relative to his own condition. Can so complex an assessment be carried out without thinking? And I wonder whether the more parsimonious explanation wouldn't be that they think.

Teaching, which has been assumed to be a uniquely human attribute, is one behavior we have observed in elephants. For example, young females coming into estrus for the first time do not seem to understand what they are supposed to do. They tend to flee from all males, rather than standing close to the highest-ranking musth male and running only from young, nonmusth males, as older females have learned to do. On numerous occasions Cynthia and I have watched a mother apparently teaching her daughter how to behave, leading her directly to a musth male and then standing with her, initiating the movements she should go through. This behavior can sometimes make it difficult to determine which of the two females in actually in estrus.

I also have always felt that elephants have a sense of themselves *as* elephants, that they see themselves as different from other species. Elephants seem to have categories in which they classify other animals. They have a particular dislike for species that are either predators or scavengers, even if those animals are not a threat to elephants or are scavengers that are not feeding on an elephant carcass. For example, if an elephant comes across a group of jackals and vultures feeding on a zebra carcass, typically it chases them off or at least shakes its head in what I would characterize as annoyance. It is as if elephants do not like the sight of blood or have a particular aversion for carnivores.

Elephants also show their dislike for animals that run around in an apparently chaotic fashion. They have a particular distaste for wildebeest during the rut, and at times they will leave what they are doing to chase a wildebeest in apparent irritation. I have found a number of rabbits and birds flattened by elephants, and Cynthia watched an elephant step on a yellow-necked spurfowl that jumped up directly under the elephant's chin, causing it to be startled. Elephant chasing and sometimes killing of "irritating" animals is different from their chasing of "despicable" animals. Whereas they generally ignore other herbivores unless they are behaving in an unacceptable fashion, they typically react to carnivores by chasing or by simply shaking their heads or flicking their trunks in their direction.

People tend to assume that if animals do have any thoughts, then they are probably confined to representing real objects and events. But elephants often interact with what I call imaginary enemies. In the cool of the evening one often finds young elephants off by themselves on the edge of a group, beating up bushes and running through the long grass as if on the attack. The elephant charges through the grass head down, picking up bits of vegetation, sticks, and branches, and tossing them in the air as it comes to an abrupt stop and, with head and tusks high, stares through wide eyes at some imaginary enemy. It then turns around and charges at the enemy in the opposite direction and around in circles. If elephants are assumed not to have the capacity for thought, are not able to remember or anticipate objects and events, then we must conclude that these young elephants are merely genetically programmed to behave in this absurd way. Personally, I believe that they are, in human terms, acting out some fantasy. As I think of my own experiences, as my thoughts have moved from fantasy back to reality, there is always a moment of self-awareness, a jolt of catching myself in the act of being someone else, somewhere else, or doing something I know that I couldn't. If we acknowledge that elephants have fantasies and an imagination, it suggests that they, too, may have comparable moments of self-awareness.

CHAPTER ➤ 18

The Games Elephants Play

When people ask what I find special about elephants, one of my answers has always been their sense of humor. To have a sense of humor, I believe one has to have not only an ability to see the funny side of a situation, to see a situation in a larger context, but also a sense of self.

The *Oxford English Dictionary* gives several definitions for the word "consciousness." One is "the state or faculty of being mentally conscious or aware of anything"; another, "the recognition by the thinking subject of its own acts or affections." These two definitions are quite distinct. The first is simply conscious perception, being aware of immediate sensory input, and perhaps also memories and anticipations. The second definition refers to a conscious awareness that one is thinking or feeling in a certain way—that is, an immediate awareness of one's own thoughts, as distinguished from the objects or activities about which one is thinking. In other words, in order to be self-aware or have a sense of self, a being must possess conscious thinking—it must recognize its own acts.

While many people feel that animals may sometimes experience conscious perception, most believe that self-awareness, being aware of one's own thoughts and feelings, is a uniquely human attribute. From my fieldwork, I believe that elephant thinking actually involves both types of consciousness. While my arguments that elephants are sometimes consciously aware that they are thinking and feeling in a certain way may not ultimately be convincing, I nonetheless want to try to make my case by describing what I call an elephant's sense of humor.

While watching elephants play games, I often get the very strong feeling that they see themselves as large, rather comical animals and, as I said earlier, that they have a sense of themselves as being different from the other species with whom they share the plains. Elephants often seem to do things for their own or others' amusement, even in as simple an act as putting on a play face, their mouths open, the corners of their lips turned up in a smile. Elephants are clowns and give many indications of being aware of it.

I have on numerous occasions watched an elephant who has finished drinking its fill raise its trunk up again and then, instead of putting it into its mouth, hold it in the air and, with the two tips squeezed together, spray water out from both sides like a fountain. The elephant appears to be amused by its ability to do this rather pointless thing with its trunk and will repeat the game, filling the trunk and spraying the water out over and over. The elephant's expression is what I would describe as self-satisfaction and self-amusement, and it is often accompanied by a particular head wag with ears alert, a movement specifically associated with an elephant's apparent consideration of its own actions in relation to some object or event. While observing the fountain display, I always wonder whether the elephant is thinking, Aren't I playing a silly game?

Elephants play many similar, apparently self-conscious games. Tuskless's family often passed through the drivers' quarters in *Ol Tukai*, walking among the houses and past the bar looking for interesting things to eat. I was there one Sunday afternoon buying a crate of beer when they visited. As the human crowd looked on from the bar, young Tom, Tania's son, ambled over to the rubbish bin to see if it contained anything tasty. He reached his trunk inside and, discovering that it was empty, settled for a game instead. He swung his trunk vigorously round and round the inside of the bin, causing a loud *bou-bou-bou-bou* sound as the base of the empty can made contact with the ground. After setting the bin spinning out of control, he walked casually off a few steps beyond it and then paused, his ears raised, listening intently, as if waiting to hear the can falling over. When it didn't fall, he reversed two steps and, without looking back, he gave the bin a big kick with his right hind leg, knocking it over with a satisfactory crash. Tom listened, the crowd cheered, and he sauntered off waggling his head and ears in that special way, as if terribly pleased with himself. It seemed apparent from this expression that not only had he wanted to make a loud noise and the bin to fall over, but that he was consciously aware of his intent and entertained by his own actions, and perhaps by the fact that he had an audience.

During Keith's study of elephant feeding behavior, he began monitoring a series of grass plots every few months to assess primary productivity, one measure of the food available for elephants. When he left it was decided that we should continue collecting this information, though Cynthia and I both disliked the task intensely. We were not ecologists, we were ethologists, and despite the importance of knowing how much herb layer biomass was available for elephants during the dry season and after the rains, we found the process of measuring grass extremely tedious.

Doing the grass plots involved finding a particular spot, marked by a bush or a rock, and unwinding a piece of string from that location to another spot a hundred meters away. This was often time-consuming because from one season to the next, we usually forgot the exact location of the rock or bush we had chosen as our original landmark. The string was marked off in five-meter intervals, and at every mark we had to count how many of the ten pins in a pin frame (used to measure herb layer biomass) we placed on the ground there were touched by blades of grass. This gave us "percent grass cover." We then had to measure the height of the grass. After we had done this twenty times, we had finished one plot and we wound up the string, put everything back in the car, and went to look for the next plot. There were twelve plots, and the exercise took two days. The hope that something unusual might happen was all that distracted us on these days.

To make it easier for us, Keith had designed "the winder," a wonderful contraption that allowed us to unwind the hundred-meter string simply by staking one end of the string in the ground and walking to the other end of the transect. To wind it up again, we held on to the handle of the devise with our left hand and wound another little handle with our right. (The alternative was to wind the string around a stick, which made the exercise even more tedious.)

One day Cynthia and I were on our fifth grass plot down by the edge of the swamp. To get to the rock that marked the beginning of the transect, we had to drive through what had once been thick young acacia but had since become a tangle of mangled thorn bushes, some still growing, some not. After we found our rock and looked around to verify there were no buffaloes or elephants nearby, I set off with the winder to clamber through the hundred meters of bush to reach the end of the transect. Having done so, I returned and we started our work, one measuring grass and the other recording data.

We had made it to about the eighty-five-meter mark when we became aware of some elephants heading in our direction, not from any particular sound they made but from the sense one gets after being around them for so many years that they are nearby. We looked up and could see a family group sauntering toward us along the edge of the swamp. We decided that it would be safer if Cynthia brought the car closer. That done, she kept an eye on the elephants from a vantage point on top of the car, while I continued to count the number of pins touched by grass stems. "It's Delia and her family, and they're coming closer," Cynthia said. "How close?"

I asked. "Thirty meters." I decided to keep going for a bit and just finished the last measurement when Cynthia said, "I think you'd really better come now; they're right here." I quickly slipped through the bushes and got into the car with her. We thought that we could simply let them pass and then get out and wind up our string, but it was not to be.

The elephants approached the string slowly, feeding as they went until suddenly, in unison, they froze in their tracks and raised their heads and ears in alarm. They had discovered string in a place where string was not supposed to be. This was cause for great consternation, because elephants do not like surprises or things to be out of order. They cautiously reached their trunks out toward the string several times before touching it. Perhaps they could tell by its smell that it had been brought there by people that they knew, because they relaxed their ears again.

It was Daniel, Delia's son, who decided that the matter should be investigated further. He stepped forward boldly, grabbed the string, and, swinging his trunk rapidly in circles, he wound the string around it. He took another step forward so that his two front feet were on one side of the length of string and his two back feet were on the other; then he proceeded to spin his body around in circles. By now, totally entangled, he simply took a big step forward, breaking the string into many pieces. Then he waggled his head in elephant self-amusement. I sighed, for that would be the end of the grass plot surveys until I could get another length of string from Nairobi and mark out the five-meter intervals all over again. Although Daniel was now bored, Delia and the other adult females moved in. Delia grabbed the end of the original section of string and followed the length of it into the bushes. We wondered what she was up to, then suddenly heard the sound of the winder being pulled through the bushes. We watched the string and its black marks moving by faster and faster until the winder came flying over the bushes toward the elephants. There was silence for a few minutes, presumably while they all carefully examined the unknown object, and then came the familiar *tick, tick, tick* of the winder being wound. The ticks got fainter as the elephants moved off through the bush with it. We never saw the winder again, though we searched through the thorns for it, and without Keith around we had to make do with a stick.

On another occasion I was out one evening in my vehicle on the southern edge of *Oltukai Orok* watching Slitear's and Estella's families. When Slitear's six-year-old daughter, Slo, play-trumpeted at me, indicating she was in the mood for a game, I couldn't resist the temptation to trumpet back at her. I licked my lips, pressed them together, and blew hard. (My play trumpets are better than my rumbles.) Slo rose to the challenge and trumpeted again. I returned her trumpet, and she raced off around a bush only to return trumpeting loudly at me. We had started a game.

Slo became more and more excited. Sucking in her cheeks, she curled her trunk under her chin and, with her head high, looked over her short tusks at me, her amber eyes wide. Then she picked up some palm fronds and threw them in the air. I responded to this challenge by tossing a piece of elephant dung at her. She reached out again and again, touching the dung with the tip of her trunk. The smell of her human friend was on the dung, which was apparently cause for great excitement. She raced around the bush again, only to reappear walking toward me backward, her tail high, the whites of her eyes showing as she watched me over her shoulder.

The commotion soon attracted other young elephants, and as Slo and I reached the edge of the Serena plains, four juveniles were running around my car trumpeting. The older adult females pretended to ignore us, but the younger females couldn't resist, and by the time we were halfway across the plains, fifteen elephants ranging from four to twelve years old were accompanying me. By now the elephants had worked themselves into a complete frenzy: Some were running wide circles around my car, while others floppy-ran several hundred meters away and then turned around and ran, trumpeting all the way back to me. When elephants are feeling particularly silly, they run about in a loose and floppy manner with their heads down and allowing their trunks and ears to flop and wiggle about. It is this behavior that we call the floppy run.

Tamsin, a young female from Slo's family, came up to the window with a tin can stuck on the end of her tusk, ran around looking in at me with her eyes rolled back, in extreme elephant excitement, and then proceeded to back into the car, trying to sit on its hood. I told her in a stern voice that this was not a good idea, and she ran off with the can still firmly stuck on the end of her tusk. Then Elizabeth ran up and put both of her slender tusks in through the window and against the car roof. I reached above my head, grabbed her tusks, and pushed her back out the window.

By now it was almost dark, and our game had to come to an end. I left the elephants at sunset and drove back to camp. Tamsin, Elizabeth, and Slo apparently remembered our foolish game for many years afterward because whenever they saw me approach in my car, they waggled their heads and ears at me, with their mouths pulled back in a smile.

One day in February 1989 I was out on the plains near the causeway with Jezebel, Joyce, and the rest of the JAs. The elephants were relaxed and sleepy, in a midmorning sort of a mood. The adult females were dozing while the youngsters played their favorite game, King of the Castle. To initiate the game, a juvenile female lay down on her side and began to wriggle her body and flop her trunk about. Babies wandered over and began to climb on top of her until there was a pile of wriggling elephants. Joshua, Joyce's teenage son, who was too old for this juvenile game but still too young and energetic to feel sleepy, came over to see what I was doing instead. I was sitting on the hood of my car and, as he approached, I rumbled softly to him. He raised his trunk high in the air and came closer. I, too, was in the mood for a game. I took off one of my rubber flip-flops, worn thin from years of use, and tossed it to him. He reached his trunk tentatively toward the shoe, relaxed and elongated it, and then swung it toward the shoe, sniffing it again and again until he decided it was safe to touch. Then he

stretched his trunk out to its full length, picked up the flip-flop, and our game began.

Joshua had not met a flip-flop before, and it required serious investigation involving prolonged touching with various parts of his body. He first stabbed it with the tip of his tusk and then used it to scratch the underside of his trunk, which made a lovely rasping sound against the ridges of thick elephant skin. Finally he put it in his mouth and chewed it gently, turning it round and round slowly with his large tongue. After several minutes of such examination, he tossed the shoe up in the air behind him. Listening carefully to where it landed, he opened and closed his eyes as if considering the interesting new sound that it had made. Then, waggling his head and ears, he reversed several steps and reached out to touch it gently with his hind foot. Having touched it carefully from all angles and with both hind feet, he stepped carefully on it, scuffed it through the dust, stepped on it again, and then, with his back legs crossed, he contemplated the flip-flop in deep elephant silence. When we humans contemplate an object, we look at it intently. Elephants, in contrast, take it in sensually. Joshua stood quietly, facing away from the flip-flop, perhaps reliving the wonderful new feelings he had experienced. Then he reversed further, reached back with his trunk, picked up the flip-flop, and started the game all over again. Finally he tossed my shoe, undertrunk, back to me. I picked it off the ground, found it relatively undamaged, and threw it back to him. We did this a couple of times until something else caught my attention, and I looked away for a minute. The next thing I knew something hard landed on my head and fell to the ground with a thud. Joshua had found a small piece of wildebeest bone and had thrown it at me with surprising accuracy. It seemed clear to me that morning that Joshua understood and was amused by our game: There we were, two species out on the plains playing catch.

One day when I was watching a group of males south of Observation Hill, Vladimir, a young male of eighteen, left the group and came over next to the car. He stood so close that one of his tusks was within a few centimeters of my window. On an impulse I reached out slowly and held onto the end of it. I don't know what I expected him to do, but I was aware that it would have been very easy for him to tusk me, had he so wished. But Vladimir just opened and closed his eyes very, very slowly. I remember feeling the cool smoothness of ivory against the palm of my hand and looking up at his long eyelashes and into his amber-colored eyes. We stayed like that for a minute or so until he began to move his body gradually in toward me, closer and closer, until his tusk was inside the car. He then allowed me to push all 4,500 kilograms of him back out until I could no longer reach him, and I let go. Then he came back toward me, again very slowly, until his right tusk touched the car mirror. The sound of ivory against glass surprised him, and he walked away.

Each time I saw Vladimir afterward I called out, "Come, Vladimir, come, come, Vlad, come," and he always walked out of his way to see me, allowing me to touch his tusks and rub my hand up and down the rough skin on his trunk.

One day, while I was engrossed in an intimate conversation with Vladimir, I noticed that we had an audience: Norah and Soila were watching us from their car on one side of us and a group of Vladimir's elephant companions were watching from the other. It struck me that the two groups may have been thinking much the same thoughts: how strange for one of us and one of them to be so interested in each other. Finally the other elephants left for the swamp and Vladimir and I interacted for a while until he, too, began to walk away. He paused a few paces beyond me and gave what sounded to me like a "Let's go" rumble, as if he were urging me to come along with him for the day. I followed, drove past him, and then waited ahead of him. Again he came over to see me, allowing me to touch his trunk and tusks, and then he walked on, paused, and seemingly called back to me. We repeated this sequence all the way to the swamp edge, a distance of close to a kilometer. I felt sure that he was telling me to join him. Eventually, still rumbling, he disappeared into the papyrus, and I was unable to follow. I asked Norah and Soila, "Did you hear him?" "Yes," they answered, "it sounded like a 'Let's go.' " Could it have been? And if so, what, I wonder, did Vladimir and his companions think of that?

CHAPTER ✦ 19

An Elephant's Empathy

BUT SEEING IT WITH MY OWN EYES I FELT THAT I WAS AT LAST SEEING
ELEPHANTS AS THEY REALLY WERE, NOT AS THEY APPEAR TO BE
WHEN THEY SUSPECT THAT HUMANS ARE WATCHING THEM.
—Vivienne de Watteville, *Speak to the Earth,* 1935

THERE IS SOMETHING EERIE and deeply moving about the reaction of a group of elephants to the death of one of their own. It is their silence that is most unsettling. The only sound is the slow blowing of air out of their trunks as they investigate their dead companion. It's as if even the birds have stopped singing. Just as unsettling is the way elephants back into their dead. Although elephants use their front legs for killing, by kneeling on their victims, they have a way of walking backward and using their sensitive hind feet surprisingly delicately for waking up their babies and touching the dead. Using their toenails and the soles of the feet, they touch the body ever so gently, circling, hovering above, touching again, as if by doing so they are obtaining information that we, with our more limited senses, can never understand. Their movements are in slow motion, and then, in silence, they may cover the dead with leaves and branches. Elephants' last rites? A wake, a death watch, the calling up of the elephant spirits? Elephants perform the same rituals around elephant bones. They approach slowly and silently, and then the touching begins, slowly, as they deliberately, carefully turn a skull over and over with their trunks, touching, hovering over the long bones with their hind feet. Watching elephants with their dead always leaves me with many stirring emotions and many real questions. Perhaps our fascination, or even our fear, is that elephants possess something that we believe only humans can have: a sense of death and therefore a sense of self, a sense of their place in nature.

I used to collect the lower jaws of elephants I had known, bringing them to the camp for aging, tagging them, and placing them, one by one, in a growing semicircle around my tent. Often elephants would enter the camp at night and carry the recently collected jaws away, and I would have to retrieve them the following morning from under the palms. One night an elephant came after I had brought back a new jaw and began to break off pieces of palm fronds from my *makuti* roof. As my roof began to shake, I called out for her to stop; I couldn't understand what she was doing. In the morning I found that the jaw had disappeared, and strewn about in its place were the broken-off bits of *makuti*. Had the elephant been trying to cover the mandible? On another occasion, I retrieved the jaw of matriarch Big Tuskless, who had died in the swamp. The following day her family arrived at camp and found her jaw; her youngest son, Butch, stood quietly over it for half an hour, gently feeling each tooth.

An interesting anecdote was related to me by Simon Trevor, one-time warden and later wildlife filmmaker resident in Tsavo National Park. A woman had come to visit him wearing several ivory bracelets, which were then still fashionable and acceptable to wear. As they were standing outside his house, Eleanor, an elephant who had been orphaned and raised by Daphne Sheldrick (now reknown for her tireless work saving baby elephants left orphaned by poachers) from a baby, approached them. As Eleanor came closer, Simon said to the woman, "I think you should put your ivory behind your back." Eleanor walked up, reached behind the woman, and, taking her hand in her trunk, carefully examined the ivory. I was intrigued by this story. Could Eleanor actually smell the ivory? And why was she so interested in it? I suggested to a fellow researcher, Barbara McKnight, who was working on elephants in Tsavo, that she might try a similar experiment if she found a piece of ivory in the park. Barbara came back from the field one day and stood near the entrance to Eleanor's *boma*, or enclosure, with the ivory hidden behind her back, where Eleanor could not see it. Eleanor walked up, reached behind Barbara's back, took the ivory into her trunk, and raised it up close to her eye to study it. Not only was Eleanor apparently able to smell something that to us has no scent, but she was interested in actually studying the piece of ivory, presumably because it had belonged to another elephant.

The response of elephants to the bones and tusks of their own species has been described and filmed many times, but it is worth recounting another one of my own experiences. Recently I was assisting a National Geographic film crew to obtain footage of elephants with elephant bones. I had often collected bones and placed them in the path of a group of elephants, in order to study their reactions, but this time was different. We were gathering the bones of Jezebel, my favorite female, and presenting them to her own family. The family approached her remains and then suddenly stopped and became silent. They neared the bones very slowly and then spent the next hour turning the skull, the jaw, and the long bones over and over. The elephants, who appeared to be in a sort of trance, neither interacted nor vocalized and seemed to focus only on the dead elephant. Jolene, Jezebel's daughter, appeared to be the most absorbed of the group. What was she thinking: This is my mother; she died in a lot of pain; life is not the same without my mother? Why would an elephant stand in silence, over the bones of its relative for an hour if it were not having some thoughts, *conscious* thoughts, and perhaps memories?

As I mentioned earlier, elephants do not like the sight of blood nor the animals that cause it, and will chase lions, jackals, or vultures off a kill. But they do not then stand over a dead zebra, slowly touching its body and burying it with dirt and vegetation, nor do they turn their trunks toward or pause to stand over the bones of a wildebeest. They reserve this behavior for their own kind and sometimes, one other species: humans. If elephants see themselves as different from other animals, is it possible that they also see humans as different from the rest of nature? And how do they measure that difference? If elephants have the capacity to think consciously, if they understand death, is there any indication that they can empathize?

Empathy is the imaginative projection of one's own consciousness into that of another living entity. When an elephant is injured, other elephants, relatives, and in some cases nonrelatives, come to its assistance, possibly showing that they understand what it means for another to feel pain. On a number of occasions I have watched elephants standing on either side of an immobilized elephant, propping it up between them. Recently Norah and Soila told me that Vladimir had contracted some kind of a disease that had left him almost crippled. They had gone to check on him for several days in a row and reported that several young males seemed to be looking after him. Albert was seen with him day after day, walking along at Vladimir's slow pace, accompanying him to the swamp, and then another young male took up the job. It was as if they understood that Vladimir was sick and needed help. But is their empathic reaction and their providing assistance simply a behavior that has been selected over the course of millions of years? Elephants who assist closely related individuals may increase their inclusive fitness and possibly their own future chance of survival and that of their offspring. But why should Albert, from the AA family, assist Vladimir, from the VAs? Were Norah and Soila witnessing empathy, or is the behavior reciprocal altruism and simply "wired in"? Most would probably conclude the latter, but if that is the case, why do elephants assist injured elephants *and* injured humans, but not, as far as we know, other species?

There are numerous accounts of the behavior of elephants around injured or dead humans, two of which I will relate here. The first was told to me by Colin Francombe, who witnessed part of the incident several years ago on Kuki Gallman's Laikipia Ranch, when Colin was manager. One of the ranch herders was out with the camels when he came upon a family of elephants. The matriarch charged and struck him to the ground with her trunk, breaking one of his legs. When the camels returned to their *boma* that evening without their herder, the alarm was raised. Early the next morning a team of trackers was sent out to search for the man. They found him propped up against a tree, a lone female elephant standing over him. The search party tried to frighten the female away, but she charged the men and chased them off. The trackers returned to the ranch headquarters to get a vehicle and assistance. Returning with them, Colin tried to force the elephant away with the vehicle, but again she charged repeatedly. Assuming that she was extremely dangerous, Colin reluctantly prepared to shoot her. As he raised his rifle to fire, the injured man shouted for him to stop. Colin resorted to shooting over the elephant's head, finally driving her far enough away for the vehicle to approach and collect the injured man. The herder related that, after the elephant had struck him, she "realized" that he could not walk and, using her trunk and front feet, had gently moved him several meters and propped him up under the shade of a tree. There she stood guard over him through the afternoon, through the night, and into the next day. Her family left her behind, but she stayed on, occasionally touching him with her trunk. When a herd of buffaloes came to drink at the trough, she left his side and chased them away. It was clear to the man that she "knew" that he was injured and took it upon herself to protect him.

The second story was told to me by a National Park Warden and friend, Simon Makulla. Simon is Maasai and grew up in the Morijoi Forest above the Siria Escarpment, overlooking the Maasai Mara Game Reserve. As a young boy he spent his days like other young boys, herding his family's livestock. He left early each morning, after a cup of tea, returning only late in the afternoon. During the day he found edible plants to keep his nagging hunger at bay. Even at the age of six or seven years old, his job often brought him into encounters with potentially dangerous animals. He regarded this as part of his education: He must learn the ways of the wild animals if he was to prove his manhood and become a warrior of any worth. On the particular day of his story, he had left home with his best friend, another boy of six or seven. They had traveled across the open plains with their cattle, stopping to explore and study things of interest on the way. Later in the day they had come to the edge of the forest, which they entered while the cattle grazed on the lush green grass in the shade of the large trees. Suddenly they came upon an elephant. It charged, and they ran in opposite directions, Simon around one side of a clump of bushes, his young friend around the other. Simon heard screams and then silence. He ran faster until suddenly, there before him was the elephant standing above the body of his young friend. Simon froze in fear, but as he stood there the elephant gently touched the still body and then, with its trunk and forefeet, slowly began to cover the boy with dirt. Having covered the body, the elephant began to break branches and twigs from the surrounding trees, placing them one by one on top of the small boy. Then the elephant stood quietly over the site. Simon watched with horror for what seemed to him like hours and then finally came to his senses and ran to the family settlement, arriving breathless and unable to speak. His state of shock and the disappearance of the other boy told the elders the story: The young boy would not be returning, he had been killed. By now it was dark and there was nothing they could do but wait until the first light of day.

Early the next morning, as the horizon turned from slate blue to pink, they left carrying their spears. They followed the path that the cattle had taken the day before, where the grass was bent over by their many footsteps and the smell of fresh dung still lingered in the cool, damp morning air, until they came to the edge of the forest. There they found the small footprints of the boys and cautiously entered the forest. In front of them the ground was bare and trampled. Tusk marks and huge footprints on the freshly plowed earth marked the spot. Suddenly they saw the elephant, standing still and silent over a pile of dirt and brush where the boy lay. They watched in silence and then, as Simon told me, they left the elephant to its vigil. Over the years I have heard many such stories of elephants covering dead or sleeping humans, and perhaps this behavior is what has led the Maasai to put vegetation into the orifices of both human and elephant skulls.

On the basis of my own observations, I have no doubt that elephants have conscious thoughts and a sense of self. Though we may never be able to collect data convincing enough for the skeptics, this subject is important not only philosophically and scientifically but also, in the context of the realities of elephant management, ethically. If elephants are conscious, thinking animals, where do we draw the line in our management policies? Is it ethically acceptable to cull entire groups of elephants, first darting them from a helicopter with scoline, which prevents them from moving but allows them to feel and leaves them aware of everything that is taking place, and then killing them later with a shot to the brain at point-blank range? Is it ethically acceptable to leave babies tied to their dead mothers as the butchering process takes place? Should elephant sport hunting be discouraged on the basis of what we know, or encouraged because it brings in revenue that can be used to "conserve the species"? Should we shoot so-called rogue or problem elephants in situations in which many of them actually end up being simply wounded or riddled with bullets only to die hours later? And if some of these practices are not acceptable, how *are* we going to deal with some of the very difficult management questions that arise?

CHAPTER ➡ 20

Amboseli Seasons

The hills were not so high, and had gentle slopes; sometimes
their flanks began to move: it was the elephants. . . .
—Romain Gary, *The Roots of Heaven,* 1958

By the late 1980s I had spent close to a decade in Amboseli and
I felt very much in tune with its cycles. The dry season followed
the rains, the dust followed the mud, the arrival of large herds of
wildebeests and zebras was soon followed by their departure. With
the seasons came changes in the behavior and grouping patterns of
the elephants. Each month a new male came into musth, and even
today if you tell me the name of a male in musth, I can tell you
the area where he's most likely to be and the length and color of
the grass. Iain, *Ilmarishari,* the grass long and green; Bad Bull, *Oltukai Orok,* in among the palm trees.

In January, when the short dry season began, the brilliant green
plains were dotted with small pink and white flowers. The elephants preferred the tortilis woodlands, where they moved in large
aggregations and fed on the long green grass that grew in the shade
of the umbrella-shaped trees. On the edge of these woodlands grew
delicately curving, sweet-smelling white flowers that carried strong
semen-scented undertones. As the elephants uprooted trunkfuls of

grass and scuffed the earth with their great feet, the air became heavy with their fragrance. If the short rains in November and December had been generous, the females' breasts were full, and the layers of fat on their babies gave their small, rounded bodies a quilted appearance. As the adult females gained weight their reproductive cycles were stimulated, and young and old males roved among the groups, carefully testing each female for signs of estrus. It was a time of plenty for the elephants, a time for breeding, and a time when Dionysius and Aristotle were in musth.

Within only a few weeks the grass turned from emerald green to golden, and the midday sun beat down from a cloudless sky, creating heat waves that shimmered across the plains and often made it difficult to distinguish, on the distant horizon, a group of elephants from a herd of wildebeests. I had learned that if the group appeared to include animals of several different sizes they were elephants, while if they were all one size they were wildebeests. I scanned the tortilis woodlands for puffs of dust, telltale signs of elephants kicking the ground, uprooting what remained of their favored wet-season grass.

By late February a steady wind swept across the plains, creating swirling dust devils that reached for the sky. My skin shriveled up like the legs of Agama lizards, but still the wind blew, and a relentless sun beat down on the parched landscape. The snow on Kilimanjaro melted away in the heat until it was reduced to a thin glacial cap. The elands, oryxes, gerenuks, ostriches, and Grant's gazelles could survive with very little water and stayed in the bushlands throughout the year, but the water-dependent migratory herbivores—the wildebeests, zebras, Thomson's gazelles, buffaloes, and elephants—returned from the *Eremito* ridge and down from the slopes of the mountain, moving back into the Amboseli basin in the thousands. The elephants began to spend the heat of the day on the edge of *Longinye* and *Oltukai Orok* swamps where the *Cynadon dactylon* grass grew in lush green swards. Cattle egrets rode on the elephants' backs or walked alongside their large feet, catching insects that had been disturbed in the long grass.

Amboseli lay covered in a blanket of fine dust, and there remained little for its animals to eat on the hard saline pans. Even the short, spiky plains grass, so well adapted to the harsh conditions, had been sapped dry by the wind and the sun. The wildebeests and zebras were listless, walking slowly through their daily routines with their heads low. They, too, moved in closer to the swamps where the daily presence of large groups of elephants had stimulated the grass to grow thicker and faster, creating a grazing arena for all. By midday *Longinye* was a sea of elephants, buffaloes, wildebeests, zebras, and rhinos. Lions lay hidden in the tall grass, waiting patiently for nightfall. The elongating shadows were a reminder to the grazing herbivores of impending danger and triggered their slow return to the relative safety of the open plains. There they waited through the night, returning to the swamps at daybreak. The elephants, too, left the swamps at sunset, retracing their steps back through the tortilis woodlands and beyond, returning to the *Acacia nubica* and *Acacia melifora* bush on the lower slopes of the mountain where their predator, man, had gone to sleep.

By the first week of March Iain, the highest-ranking of the males, began to show signs of coming into musth. He returned from his sojourn south of the border, making his appearance in the predominantly female areas of the *Ilmarishari* woodlands, *Longinye* and *Enkongo Narok*. Although he walked alone, he held his head high, and his temporal glands were slightly swollen, with a tiny black dot of secretion visible at their orifice. Within a couple of weeks he had moved into the midst of the female groups and was in full musth, dribbling and secreting profusely. Dionysius and Aristotle, too, were in musth, but they had now lost condition and were careful to avoid Iain.

The air was heavy and close, and we found ourselves searching the skies anxiously: Would the long rains come this year, or would the grass shrivel up and die, leaving the animals to starve again? There is a Swahili saying, *Dalili ya mvua ni mawingu,* "The sign of rain is the clouds." Ours were the "marching clouds," the small, puffy clouds that came from the direction of the Chyulu Hills, gradually gathering into the angry slate-blue skies that brought rain to Amboseli. Although I was careful not to point at these promising skies, weeks would pass with the marching clouds sailing softly by only to develop into thunderheads on the other side of the mountain. Then, finally, the rain clouds came and towered over Amboseli, and as the late-afternoon sun descended below them, the yellow acacias shone golden against the gray sky.

The army ants, *siafu,* suddenly appeared, making their way through camp in long angry red lines. We viewed their arrival as a mixed blessing; they were a signal that the rains were imminent, but the warriors could badly bite an unsuspecting person stepping on their trails. They invaded the kitchen and the *choo,* and we sometimes woke in the night to hear the soft sound of millions of tiny legs circling our tents, forcing us to venture into the dark to pour ash over their trails.

We waited, too, for the full moon, for with it, it was said, came a change in the weather. The wind, the clouds, the moon, and the animals, they were all our *dalili ya mvua,* and we monitored them closely as we waited each day for the rains to come.

In late March the first raindrops fell on the deep powder, each raising a small puff of dust. With the first rainstorm Amboseli was transformed. The palms and the acacias were washed, and everything had a fresh, green appearance, as if the land had been given a new coat of paint. The dry dust that had smarted our nostrils for months was replaced with a rich, damp, earthy smell. We could breathe deeply now.

The rainwater filled up the pans on the plains and poured down onto the roads, and Amboseli changed from a flat, barren plain into a series of islands. Frogs appeared from nowhere, and catfish found their way into the newly created swamps and lakes. In places the roads became rivers, and I had to draw heavily on memory to avoid the huge potholes where water would seep in through the doors of my car. The main road through Amboseli had been graded so many times that the ditches originally designed to drain the water from the road and onto the plains now did the reverse. At the

Mtaro ya Mungai, on the way to *Ol Tukai,* water came halfway up the side of the car. The alternative, which most cars chose, was to risk the deep mud on either side of the road, and we grew accustomed to waking to the whining sounds of straining engines and spinning wheels.

Sometimes it rained all night, and we lay in our tents listening to the pitter-patter of raindrops on the *makuti* roof, waiting for the occasional drop to escape through the layers of palm and fall with a *ping* on the canvas. It was comforting to lie warm and dry inside the tent knowing how wet and muddy it was outside. After such nights we would wake up to find that it had snowed on Kilimanjaro, the snowfall often descending down the slopes of Kibo and its companion peak, Mawenzi, to meet in fresh powder on the saddle.

Within days of the first good rains, green shoots of grass forced their way to the surface, but the wildebeest and zebras had long since left in search of more nutritious pastures outside the basin. Only the high-ranking male wildebeests remained behind to guard their territories. They looked so lonely lying out on the bare pans that we couldn't help but wonder why they bothered to guard such an uninspiring piece of property. But these small plots were key to their reproductive success; they were areas the females would pass through on their way to the swamp during the rutting period in June, and if they abandoned them now, others would quickly take their place.

As the rains continued into April, the deep volcanic powder of the Amboseli basin was transformed into a rich slimy goo. It was crucial at this time of year to "know your soils," as Cynthia used to say. It took experience, which meant getting stuck, to learn the Amboseli soils well. I got stuck in the Amboseli mud on many occasions, but the worst time was when Meloimyiet, Sandy Andelman, and I went out one night to rescue Cyn, who had failed to return to camp. We spotted her lights a kilometer off and, in attempting to find a route in to retrieve her, we, too, became stuck in the mud up to our axles. After four hours of digging in the mud, jacking the car up, and digging some more, we abandoned our efforts and decided, against Meloimyiet's better judgment, to walk to Cyn in the moonlight. Barefoot, half naked, and knee deep in mud, we had to chase off wild dogs with rocks and sticks. With time I learned to be particularly careful of the soil where either the elephant grass, *Sporobolis consimilus,* or the salt bush, *Suaeda monoica,* grew.

During the rains the elephants walked especially slowly, carefully placing their large feet so that they didn't slide in the smooth, slippery mud. In the soft gray soil where the elephant grass grew, their footprints left huge holes in the mud that lasted through the long dry season, making off-road driving extremely uncomfortable. Like the other herbivores, the elephants had once abandoned Amboseli during the rains and marched north to the Selengei, but that was before the poaching began. Now they remained behind, moving out at night only a few kilometers from the park boundaries.

By early May the long rains had ended. The elephants were fat and happy and moved in huge social aggregations of several hundred west toward the lake, east through *Olodo Are,* and south beyond the tortilis woodlands. The females were in estrus and the males in musth. It was a time when juveniles spent the evenings beating up bushes, rolling logs with their hind feet, tossing sticks high in the air, or charging back and forth through the long grass, peering out at imagined enemies over their short splayed tusks. Whole families "floppy-ran" across the pan, their "play trumpets" shattering the calm of the evening, the dust churned up by their scuffing feet turning pink in the sunset. Usually dignified adult females become soft and wiggly, their great heads swinging, their ears flapping, and their trunks flopping, as if their entire bodies had turned to rubber.

Wildflowers bloomed in Amboseli, and the grass was long and green and bent over with the weight of each inflorescence. The milk of the Maasai cows was thick with cream, and we drank smoky *maziwa lala,* milk that has slept, from the calabash. May was a beautiful month, but it heralded the beginning of the long dry season.

By the end of May, Bad Bull, the horror, was in musth, and he took command during the gloomy months of June, July, and August. The nights and early mornings were cold, and we could see our breath when we got up in the morning. The elephants moved in smaller groups, spending most of their days in and around *Oltukai Orok*. During the cold season a blanket of cloud descended upon Amboseli and the air became still and quiet, as if we were listening to the world through earmuffs. The colors faded into monotones, and the mountain disappeared into the haze.

The September wind blew the haze from Amboseli but brought with it a return of the dust. Bad Bull dropped out of musth to be replaced by Oloitipitip and Richard. As the long dry season progressed, the morning dust devils developed into suffocating dust storms by late afternoon. We rolled up the windows of our cars and zipped up our tents. The air was dry and the acrid dust smarted; we gritted our teeth, closed our eyes, and cursed.

As the dust and wind continued through October, depression set in. Once again we began to search the skies anxiously for our *dalili ya mvua*. The females and calves moved in small family groups, and most of the males returned to their bull areas, where they fed on the bark of acacias. By early November there were no females in estrus, no males in musth. Finally in mid-November the rains broke; it would be another six weeks before the females would begin to cycle again, before the plains would be dotted with pink and white flowers and the tortilis woodlands filled with semen-scented undertones.

Amboseli's nights were as impressive as its seasons. In camp we were more aware of the stars than we were in the city, and we lived by the phases of the moon. When the African new moon lay on her back at nightfall, the Milky Way was the only light in the night sky, and we had to guess which species visited camp by the sounds they made in the darkness: the sound of a man sawing— a leopard's rasping call; the heavy footsteps of a man—a lion passing close to the tent; heavy breathing by the edge of the canvas—a porcupine; a *ping* on the roof of the tent—a bushbaby bouncing after insects. We knew the difference between the *swish-swish* of a hippo grazing, the desperate chomping of a wildebeest, and the rhythmic *rip-ripping* of grass of an elephant. We often woke in the small hours of the morning to the sound of deep, slow snoring. As the earth became light we saw the large gray mounds of sleeping elephants by our tents. One by one they stood up, mothers waking their babies with a gentle touch of their hind feet.

When the moon rose over the plains as a gigantic orange ball, the snows of Kilimanjaro and the volcanic ash of Amboseli glistened silver, its light and the palm fronds around my tent glittered like tinsel on a Christmas tree. If camp was beautiful at full moon, spending the night with the elephants during this phase was magical. It was as if I had been granted the rights of one of the privileged few, given a ticket to join the spirits of another world. I watched in awe the huge, gently swaying shapes casting shadows across the brilliant plains, gliding slow and silent, as if they had something to hide, like dhows laden with ivory, slipping through the reef *mlango* (door) in the moonlight. The only sound was the soft crunching of crusty saline soils under huge, wrinkled feet.

I had once watched a family of elephants walking single file in the silvery radiance, swinging their trunks in unison as if in time with some infrasonic rhythm. As their trunks scraped back and forth across the bare soil, I couldn't help but think that now, under the cover of darkness, they could be themselves and revel in their love of music.

There is a time in the small hours of the morning when the wind stops and the Amboseli plains fall still. Then the elephants lie down, one by one. Whole families of elephants recline side by side, their deep rhythmic breathing the only sound in the night. As they rest I step down from my car and walk among them.

Being so closely attuned to the world around me is stirring. Through Amboseli, the elephants, and my Maasai friends, I had experienced so much of what I like to call the essence of life. The glint of an elephant's tusk in the moonlight; the feeling on my hands, in the darkness, of the rough walls made of cow dung; the pungent smell of the trail of a musth male; the smell of wet earth after the first rains. I loved the sensual side of Amboseli. But it was, perhaps, just this passionate, impulsive side of my personality that colored some of the troubles that lay ahead. For there was a dark side to my life in Amboseli, as well. Like the elephants I studied, I was getting older, but while they led normal elephant lives, mine continued to be unusual, relatively solitary, and not quite what I had hoped it to be. By the end of my time in Amboseli I could no longer take any pleasure from the sights and smells and feelings that had once been so evocative.

PART→FOUR

A REMOVAL OF VEILS
1984–1989

CHAPTER ⇒ 21

The Incident on the Ngongs

To find but one other who has seen and felt as we have
is assurance enough to brace us against
a world of unbelief.
—Vivienne de Watteville, *Speak to the Earth*, 1935

ONE INCIDENT THAT OCCURRED near Nairobi in July 1984 colored my attitude toward life for many years, and I still find it disturbing how a single event, fixed in the passage of time, can change one's perspective so unforgivingly. I have often wondered whether another woman, perhaps one less emotional, less introspective, or possibly stronger than I, might have been able to brush the misfortune aside. Nevertheless, after the incident on top of the Ngong Hills I became a very angry young woman, my enthusiasm and passion for life was deeply affected, and I began to view the world, and particularly men, as primarily hostile.

Perhaps watching elephants for long hours, immersed in my own thoughts, provided me with more time to dwell on what had happened than was good for me. Often I tried to encourage myself, by arguing that to survive in Africa, one had to make a game out of conquering adversity. *Unashinda au unashindwa,* I used to say to myself: "You conquered or you were conquered." During the late 1980s, as I dealt with the emotional aftermath of the incident, I found that making a story out of things gone wrong—to try to laugh at my own mistakes or misfortunes—always seemed to help. And out watching elephants, in a detached way, I often began the story as I imagined I would tell it some day:

Considering all that had happened, it was ironic that when one of my fellow researcher friends had first come to Africa all those years ago, we had sat together in my tent and I had read aloud to her these first few lines from Paul Scott's *Raj Quartet*:

Imagine, then, a flat landscape, dark for the moment, but even so conveying
to a girl running in the still deeper shadows cast by the wall of the Bibighar
Gardens an idea of immensity, of distance . . .

Even then the words had seemed hauntingly familiar, as if it had been we ourselves running through the Bibighar Gardens. That was almost eight years ago. Now she was leaving and I was staying. It was the end of her study, the end of her fear.

So, imagine, then, an immense landscape and two young women half running, half walking back from the top of the Ngong Hills. Covered in mud and bruises, numb with cold and shock. Their clothes torn, their naïveté gone. In material terms what we had lost was nothing: fifty shillings, our lunch of four *samusas* and two *mandasi,* and a brass bracelet; a gold ring stayed on my finger untouched, and the fear of bad spirits and a physical fight had left my father's Zeiss binoculars still hanging around my neck.

We had been sitting there, with our map, at the top of the southernmost peak of the Ngong Hills, looking down the steep sides into the unforgiving land that was the Great Rift Valley. From there, though we could see 200 kilometers in every direction, we looked south into *Maasaini,* the land and the people we knew best, home. Stretched out in the sun, contemplating the view, we were planning our next safari. A walking trip over the Nguruman Escarpment, or perhaps to *Oldoinyo Shombole,* another adventure. It

was exciting to live life the way we did: two women alone in the
bush, two women who had chosen to be different. Then, suddenly,
the three men were there.

We held each other's eyes for a long time, giving each other the
courage that we knew we would need. We both knew intuitively
what was going to happen and that it was inevitable. There were
three of them, they were apparently armed, and it was five kilometers
to the car and the closest help. Our time, it seemed, had come. This
was the situation to which our mothers had alluded. This was what
we had feared since we were young. This was the reason that we
didn't walk alone at night, why we looked over our shoulders when
footsteps approached. It was the noise under the bed, a rustling in the
bushes, a key turning in the lock, a dark shadow at the window. The
bicycles on the path, the boats on the lake, a car engine on the track
into camp, voices in the night. The situation felt familiar, as if we had
lived with its threat all of our lives.

We held each other's gaze as we were forced down the hill,
forced down to where we would not be seen. "It is going to happen," I said. A statement of fact. "Yes," she answered.

Stop talking, or we'll kill you!

Don't turn around, or we'll kill you!

Later, she found it ironic that the eagles of the Ngongs had been
watching us. What had gone through their heads as they circled
above watching the scene below? Lying on her back she had seen
them, when the blindfold came off, and the men were too involved
to remember to protect their identities. As she lay on the bare
ground watching the birds watching her, she repeated the comforting words from *Out of Africa* to herself:

If I know a song of Africa,
does Africa know a song of me?
Will the eagles of the Ngongs look out for me?

Yes, a thousand times, yes, one wants to believe, but who knows
the heart and soul of Africa?

Later, choking back the tears, she would say to me, "They just watched us! They didn't do anything!" "Who?" I had asked, confused. "The eagles!"

Betrayed by Africa and betrayed by the eagles. But, perhaps, like the Africans themselves, the eagles were used to trouble. It wasn't strange; these things happened every day. A Swahili proverb: *Sikujali taabu, kupata siajabu,* I don't mind trouble, it is not unusual. And in any case, what could the eagles have done? It was, after all, another *Shauri ya Mungu*—God's affair. Like so many events in Africa, it was simply Fate.

And so we had held each other's eyes. The eagles had watched us and we had watched each other with a look neither of us would ever forget. Eyes full of sadness, eyes that finally understood, eyes that would never see the world the same way again.

Later, after we had retraced our steps up and down the four hills back to the car, we had driven straight to the police station in Ngong town. We had staggered up the path to its entrance, past a sea of black faces that stared but expressed nothing. These things happened every day. The policemen had laughed when we arrived, and they had laughed as they took down our statements, detail by detail. After all, what had these foolish white women been doing up on the hills alone without men to protect them?

Later, as the doctor examined us, she said, "It is not your fault. You must not feel guilty or ashamed." *Pudenda:* Female genitalia. It seemed such a lovely word, pretty sounding: *pudendum, pudenda,* that of which one ought to feel ashamed. We had our examinations. There was no physical harm done.

Months later a male doctor asked me, "Have you talked to someone? You mustn't let the anger build up." The anger? I didn't understand then. "I am not angry," I replied.

The anger. It was said as if anger were bound to happen. *You must not let the anger build up,* he had said. *Why not?* I wondered. Was it because too many angry women would be a powerful force? Too many angry women could upset the balance, change the equation? Upset the male club of intimidation?

The male doctor had implied that I was not supposed to be angry with men. At the time it had seemed to me that he meant it was bad to be angry. Was this because the doctor felt that what had happened was somehow not the men's fault, and in any case, not all men were bad? Or should we be grateful that we have men to protect us from other men?

You must not let the anger build up. The more I thought about what he had said, the angrier I became. It was suddenly so clear to me that the incident on the Ngongs was just one end of a continuum of male domination. The intimidation had always been there, subtle but ever present. For me there was life before the Ngongs and life after, but there was no going back.

An elephant bellowed and I was jolted back to the present. No, I hadn't been paying attention and I had missed an important series of interactions. But it was what went through my mind so often. *The incident,* as we called it. We were different now, different from other people. It had happened four years earlier and yet it followed us around wherever we went and with whomever we met. It followed us like some despised ex-lover, like some unshakable disease. During the late 1980s I interpreted everything through a wash of anger. Though the anger helped me to cope, without it I might have made different decisions and perhaps avoided much of the unpleasantness that was to occur.

CHAPTER ➤ 22

Fear and Loathing in Amboseli

NIMEONA MENGI, SIGUTUKI.
I HAVE SEEN MUCH, I AM NO LONGER SURPRISED BY ANYTHING.
—A Swahili Proverb

ON LAZY AFTERNOONS IN camp prior to the incident, Pili used to tell me that she was writing a book, a novel, but one based on fact. It was to be a story of life in the bush based on real characters and real events. It would be a story of the Machiavellian twists in the relationships between foreign researchers and their African hosts, and between the various researchers themselves. The book, she said, would be called *Fear and Loathing in Mapagoro*.

Mapagoro was a village outside of Ruaha National Park where Pili and her colleagues had been stranded for two days when their vehicle broke down. She had started her research career at Jane Goodall's chimpanzee study site, Gombe Stream, in the mid-1970s, and had been there when a group of Zairians had come across the lake in the night and kidnapped four American researchers, holding them for ransom. She herself had escaped being captured by running through the bush in the night. The event gave those who had been there a sense of solidarity, a sense, perhaps, of belonging to a group that understood life in a certain way: the Gombe crowd. I came to know many of them while I was studying at Cambridge. After the hostage episode, Gombe was considered unsafe for young foreigners, and Pili was sent south, through Mapagoro, to Ruaha to study baboons. There, she told me, on her first day, she walked into an outhouse and found a man with a noose around his neck, swinging slowly over the long drop. She was informed by the local people simply that he belonged to a depressed tribe.

To many of us working in remote areas of Africa, events such as these came to be viewed as normal, almost expected. The Maasai have a saying, *Epwoonu illimot anaa enkolongi,* "The events follow one another like days." Strange and difficult times made for good stories, and the collection of stories and storytelling was an important component of life. Events like these also helped to develop character and a tenacity that was crucial for surviving alone in the bush.

Pili knew, then, about fear, and perhaps her interactions with some individuals in the small-town atmosphere of Amboseli had taught her something about loathing. Now she was going to write about it. I liked her choice of title, and after the incident on the Ngongs and Pili and Keith's departure I used to repeat it to myself at appropriate times, substituting the word Amboseli for Mapagoro: *Fear and Loathing in Amboseli.* It didn't have the same ring as Pili's title, but I understood the sentiments well and thought that I, too, would someday write about them.

In camp we had joked about living with just a "thin wall of canvas" between us and the wild animals. But, in fact, it was never wild animals that caused us any fear, it was people. Fear of people and the power that they had over our lives became almost a natural state. While at times after the Ngongs incident I lived in fear for my safety, more often I feared that, at the mere whim of some official, I could lose access to what was most important in my life: the elephants, the camp, living under the stars at the base of Kilimanjaro, being a part of Africa. I was, after all, only a visitor whose presence was tolerated from visa to visa, dependent on the goodwill of a series of underpaid bureaucrats.

Often I felt that I lived under a state of siege, that anything could happen at any time. This mentality was not just paranoia; the worries were real. One warden had threatened me with expulsion from the park for taking a visiting Cambridge professor out to watch the elephants without prior permission, and he had forbidden any of the Amboseli researchers from spending time in the field with scientists from a separate Amboseli project. The Office of the President and Research Council officials had once delayed my research clearance for ten months, and on another occasion I received a warning that research officials within the Wildlife Conservation and Management Department were complaining about the company I kept. On numerous evenings drunk rangers pointed loaded guns through the car window at me for merely requesting permission to pass through the *Ol Tukai* gate after the 6:30 P.M. closure. During one period bandits armed with automatic weapons ambushed cars on the road between Amboseli and Nairobi, and there was a night when bandits, one of whom apparently was recognized as a ranger, sprayed the lodge reception area with bullets. Over time, the amusement I had once felt at the seemingly illogical rules and regulations gave way to a feeling of persecution, and I began to feel increasingly threatened and insecure in Amboseli.

In fact, part of my situation was the familiar conflict between park managers and researchers, an uneasy relationship anywhere in the world. Researchers, with their detailed knowledge of a subject, always feel that they know better than a manager—or anyone else, for that matter—and frequently come to feel that bureaucrats are incompetent. Managers, on the other hand, view the issues and the problems of their park from a much broader perspective and tend to find researchers impractical, arrogant, and irritating.

The conflict seemed to have been exacerbated in Kenya during the late 1980s, and deep suspicion about possible sinister motives had developed between the two camps. Ten years earlier East Africa, and particularly Kenya, had been a mecca for a new breed of field biologist, the behavioral ecologist, and during the 1980s they continued to flock to Kenya, looking for potentially exciting socially complex mammals. I was one such researcher. Many of us were foreign students from prestigious American and European universities, determined to make our names with another trendy study. Kenyan officials perceived us as there for purely selfish reasons, caring very little whether our work contributed to the better management of the parks we worked in or whether we assisted in the scientific training of young Africans. They claimed that we left the country without depositing copies of our reports and published papers with the appropriate authorities. In many cases, we were seen as bringing little benefit to Kenya and abusing the privileges granted us. We were, in short, considered parasites, and it seemed to me that we were frequently dealt with as such.

In reality, the majority of researchers were genuinely dedicated and concerned conservationists; however, the behavior of a number of notorious individuals had long-term repercussions for the rest of us. Some scientists carried out research without permission or, having obtained permission for a specific study, researched an unrelated and sensitive topic instead. Other researchers carried on simultaneous businesses without obtaining necessary work permits, many took on foreign volunteers as research assistants instead of training Kenyans, and a large number failed to deposit their published papers with the authorities as required. As a result of their behavior, additional rules and regulations were established that made doing research in Kenya increasingly difficult and unpleasant. In the end, the Research Council, which approved projects, achieved what seemed to be its objective, and by the late 1980s very few foreigners were choosing to do wildlife research in Kenya. Consequently the country lost the edge that it once had for being the site of some of the best behavioral work in the world. Many of us persevered, but the stress of the late 1980s took its toll.

In Amboseli Cynthia and I often began to feel that we were unwelcome and that our work was not appreciated. We were known and respected internationally for our elephant studies, and yet, at the very site of our research, we were being viewed more and more as just another management problem. Instead of capitalizing on our reputation and using it to benefit Kenya's important tourist trade, we were harassed. What kept both Cynthia and me from giving up were the elephants and their survival and Kenya and its future. The elephants were not just another species to study, they were animals to whom we had dedicated our lives, and in the late 1980s they desperately needed our voice. Kenya was not just a place we had come to visit, it was our home—and we cared deeply about its fortune.

But why, if only a relatively small proportion of researchers misbehaved, was the mistrust and misunderstanding so severe? Like every conflict, there were two sides to the story, but I believe that much of the problem actually lay with the Wildlife Conservation and Management Department, the body responsible for Kenya's wildlife both within and beyond the parks and reserves. By the mid-1980s park managers were working for a system that was in a sad state of decay.

Prior to 1975 there had been two wildlife authorities in the country: the efficient and prestigious Kenya National Parks and what became the less efficient and corrupt Game Department. The government-run Game Department was responsible for wildlife activities outside of national parks, which primarily involved the granting of hunting licenses and the control of problem animals. In both areas of its operation there was considerable scope for the exploitation of elephants and other wildlife. The Game Department was also responsible for the Ivory Room in Mombasa, where the country's ivory stockpiles were stored, and for the export of considerable quantities of ivory. During the 1970s the department became deeply involved in the export of ivory from illegally killed elephants, and during this period Kenya's elephant population plummeted from an estimated 167,000 in 1973 to some 63,000 by 1979.

In 1975 it was decided that the two authorities should be merged to form one government department under the Ministry of Tourism and Wildlife; thus, the Wildlife Conservation and Management Department (WCMD) was formed. Revenue from the national parks now went to the government's central treasury instead of going directly back into the parks. With competition from the needs of the country's growing health, education, and other development programs, wildlife was not considered a priority. Without realizing it, the government, through neglect and outright abuse, was quickly destroying one of Kenya's most lucrative resources. By the late 1980s, the national parks had very few remaining serviceable vehicles and very little fuel to run them, and there was no money to repair the roads or the sign boards. Also, the parks' staff were underpaid, they wore old uniforms, they carried malfunctioning World War I and II 303 rifles, and they lacked leadership and morale.

Most wardens and rangers were deeply committed to conservation and carried out their jobs admirably under very difficult circumstances. But aware of their inability to run the parks efficiently and effectively, some managers became hostile and defensive in reaction to what they perceived as negative judgments of their performance by researchers. From their perspective, researchers were spies. Ironically, in some respects, this perception was correct. As biologists, we had been trained well to watch and to listen. Not only were we the eyes and ears of the park, we also were part of the local community, and we saw and heard much of what went on. We knew, for example, that an assistant warden was suspected of having shot the dead elephant found in *Kitirua* with its tusks missing and that his only punishment was to be transferred to another park; we knew that certain individuals were siphoning off huge amounts of entry fees at the park gates; we knew that some rangers were supplementing their salaries by illegally trading cattle across the border; we knew that others were shooting wildebeest for meat; we knew that some parks staff were cutting down trees inside the park boundaries to make charcoal; and we suspected that if we hadn't been there, the Amboseli elephants would not have survived the 1980s intact. We knew many things that the park management preferred us not to know, and by the latter part of the decade paranoia had developed on both sides. But if the situation in Amboseli was not particularly good, what was happening in some of the other parks was horrifying. With the price of ivory running at over U.S. $250 per kilogram on the international market, a significant number of individuals in the WCMD became directly involved in the killing of elephants and in the smuggling of ivory.

An incident that took place in 1987 gives an example of the mistrust that had developed in Amboseli. I had gone to park headquarters to collect data from Amboseli's Ivory Record Book on tusk recovery dates, tusk weights, and causes of elephant mortality. By comparing the information in it with the Amboseli Elephant Project mortality records, I hoped to get an indication of the proportion of tusks that were actually recovered from the total number of elephants we knew had died. I had planned to present this information at a meeting of the African Elephant Specialist Group (elephant experts selected by the Species Survival Commission of the International Union for the Conservation of Nature). When I asked to examine the book, an assistant warden told me that the information was secret and that I could not have access to it. In the paranoia of the times, she probably assumed that I was trying to find out whether wildlife personnel had been poaching elephants or smuggling ivory. While those possibilities had, in fact, not occurred to me, by being prevented from seeing a book that I had consulted many times in the past I couldn't help but wonder whether they had something to hide. After that incident the assistant warden went out of her way to be difficult, stopping me for various alleged infractions of park regulations and finally refusing to speak to me.

At about that time Amboseli received yet another new warden. The WCMD had a policy of not firing wardens but merely transferring them from park to park. By then we had been through as many wardens as years I had been in Amboseli. In looks the new warden reminded me of Idi Amin, and my first impressions of him left me feeling distinctly uneasy.

Soon after the warden arrived, two park officials—a man and a woman—came to visit me in camp. The woman was in a friendly mood, chatting away pleasantly. Then, as they prepared to leave, they both insisted that I come on patrol with them because, they said, I "might learn something." I could think of no good reason why I shouldn't join them, so off I went. Once out in the vehicle, the man began to threaten me subtly: If I did not join them for dinner that night, there might be consequences for my research. I should come with him, and he would drive me back to camp after dinner. I argued that I had a number of camp errands still to do and that another day would be better, but he continued to press me. I wanted to get off to a good start with the new administration, and although my intuition was making me question the truth of their motives, I convinced myself that I was merely being invited for a meal. The woman must have read the suspicion in my eyes for she abruptly said, "Let her come in her own car." Although her comment seemed odd, as if they had a prearranged plan for me, I reasoned that nothing could happen if I had the independence of my own transportation, and I reluctantly agreed to join them.

Dinner passed uncomfortably, if uneventfully, except that as the evening wore on, they began to insist that it was too late and too dangerous for me to drive back to camp on my own and that it would be better if I stayed the night. Then the pair excused themselves momentarily, disappearing down the long passageway to the other end of the house, leaving me alone in the sitting room to contemplate how I could leave without appearing rude or getting into an argument. Suddenly the generator lights went out, and I found myself fumbling for my car keys in the dark. My body was now screaming to me that something was very wrong, and as the woman reappeared in the darkness with a candle I heard her say, "He loves you very much. He loves you very much. He is waiting for you in the room at the end of the hall." After she disappeared back down the hall I ran for the door and turned the key in the lock, only to find that I could not open it. I fled to the kitchen door and, in a shaft of moonlight, could just make out that it had two locks. I ran back to the main entrance and turned the second lock as I heard the man's footsteps coming down the darkened passageway toward me. I escaped, but after the Ngong incident, the visit had shaken me deeply. I had not heeded my intuition; never again would I ignore it.

The man arrived in camp the following day. In front of Norah he looked me straight in the eyes and said in a measured tone, "I will keep coming until I get what I want." After that I hid whenever cars drove into camp, and I went to sleep at night listening anxiously to the sounds of vehicles passing on the main road, imagining how I would run through the bush in the night if one sought me out.

The antagonism between park management and researchers was exacerbated further by the government's increasing repression and xenophobia. At some stage in 1988 some foreign journalists apparently entered the country on tourist visas and wrote an unsavory report about Kenya's human rights record. President Daniel arap Moi responded by warning the country to beware of spies posing as foreign researchers. As a direct result, the Office of Internal Security halted research being undertaken by foreigners. The ban was in effect for several months before it was lifted, but a new restriction was put in place: So that our movements could be monitored, all foreign researchers in Amboseli were required to report to the District Commissioner's Office in Kajiado once each week. Kajiado is a three-hour drive from Amboseli. As it happened, this requirement was never effected, but the new warden took to monitoring our activities with a vengeance.

Under the warden's new rules, we were not allowed to have any guests visit us in the park without receiving written permission from him. The request had to be made on a special form and submitted two weeks in advance, in triplicate. Permission was usually refused. Since I didn't want to be placed in any situation in which I would be beholden to him, I simply didn't have any friends come to visit. When we inquired why the request was necessary, we were told that instructions had come "from above." When pressed, he told us, with some authority, that some Bulgarians recently had escaped across the border into Tanzania. We were informed that the Central Intelligence Division (CID) had reason to believe that they had found their way to Tanzania through Amboseli National Park, and the CID had concluded that the elephant researchers must have helped them. That Bulgarians had any reason for trying to escape from Kenya was hard enough to believe, but that they needed help from us to find their way south across an open plain with Kilimanjaro looming above them was ludicrous.

By 1988 I had had enough of the ridiculous rules and regulations and I was beginning to realize just how seriously the stress of physical and psychological threats had affected my emotional state. Africa and the animals were no longer enough to quell the loneliness I was feeling, and my isolation only added to my feelings of vulnerability and persecution. I was in my early thirties, often alone in camp, and I was desperately looking for someone to provide some stability and security in my life again. I recognized that if I didn't change my lifestyle and my outlook, I would never realize my dreams of marrying and raising a family. But at this point another situation rose that claimed my attention. While I sat recording vocalizations uncovering the secret life of elephants, a war was raging beyond the borders of Amboseli. Across much of the country, bandits with automatic weapons were slaughtering elephants for their ivory. How could I remain among these content, protected elephants when elsewhere entire families were being gunned down for profit? I knew that for my peace of mind, I could not stay on in Amboseli much longer.

CHAPTER ➜ 23

The Elephants' Graveyard

THIRTY THOUSAND ELEPHANTS: THREE HUNDRED TONS OF IVORY,
IF THAT. AND AS THE AIM OF GOOD GOVERNMENT IS TO INCREASE
PRODUCTION, I'M SURE THAT THIS YEAR WE SHALL DO BETTER. . . .
WITH A LITTLE GOOD WILL, WE SHALL CERTAINLY MANAGE, TAKING
AFRICA AS A WHOLE, TO KILL A HUNDRED THOUSAND
ELEPHANTS A YEAR, AND SO ON TILL THE CEILING IS REACHED,
IF I MAY PUT IT THAT WAY. IT WILL THEN BE NECESSARY
TO PASS ON TO OTHER SPECIES. OURS, I SUGGEST.
—Romain Gary, *The Roots of Heaven*, 1958

FROM THE EARLY DAYS in Amboseli, elephant bones had become landmarks. Reference points on a featureless plain, they gave special significance to an otherwise nondescript bush. As I drove through the park day after day, I silently acknowledged these places and remembered the elephants. There was the lone acacia on the way to the dry lake where Sara had died after a long illness; the section of *Longinye* swamp where Big Tuskless had fallen; the secret glade of palms deep in *Oltukai Orok* where Norah and I had discovered Priscilla's skull and tusks; the place in *Olodo Are* where I watched Polly fall and die; the pan beyond the camp where Tonie stood for two days guarding her stillborn baby. These elephants had died a natural death, but many of the others now memorialized by the bones had perished at the hands of humans. There was the spot in the thick Suaeda bush at *Njiri* where Oriel died of deep spear wounds, and not a kilometer away, under a large *Acacia tortilis,* was the place where Harriet finally fell after tossing her young Maasai assailant thirty feet through the air. There was the spot in *Oltukai Orok* where Flop Ear succumbed to septicemia caused by the spear of a young Maasai, and near the waterholes east of *Ilmarishari* was where four-year-old Rex had become a pin cushion, the day's entertainment for some brave young warriors. At a Maasai well, near the Tanzanian border, I found two dead babies floating on the water's surface, covered in a living mass of maggots, scrape marks in the earth the only clues to what might have been their fate. There was Emily's grave by the lodge rubbish pit, her stomach contents—bottle caps, glass, and batteries—revealing the cause of her untimely death. In the elephant grass, along the eastern side of *Longinye* swamp, I remember the young male with a broken leg who was euthanized by rangers, the "mercy killing" taking close to half an hour. At *Kitirua,* deep in the thick regenerating acacia, we discovered another young male lying camouflaged under a pile of brush, his faced hacked off by ivory poachers. Each time I passed these places I remembered how I had found the elephants whose bones now lay scattered there.

There is something so grand about the life of an elephant, its great size, strength, and age, that in death its loss is equally monumental. To have taken so many years and eaten so many trees, to have become so big; to have roamed the earth as King of Beasts and then to have collapsed in a piece of rotting flesh is tragic and so seemingly wasteful of life.

The stench of rotting elephant is unforgettable, and coming upon it unexpectedly still sends my heart and mind racing. Where is it? Who is it? To what horrific death did he or she succumb this time? I lean my head out of the car window, willing my nose to lead me to the carcass, acutely aware of the limitations of my human sense of smell. I am aware, too, that if I can detect the stench from a hundred meters away, all of Amboseli's elephants must be able to smell it, too.

In the Amboseli that I knew some elephants were speared by

young Maasai warriors for fun or to prove their bravery, while still others were killed as a form of political protest, but very few were killed for their ivory. The Amboseli elephants were lucky. Not so the other hundreds of thousands of elephants across the continent. From the inception of the Amboseli Elephant Project through to the late 1980s, Africa's elephants had declined from more than 1.3 million to 650,000, and Kenya's population from some 167,000 to less than 25,000. By 1987 elephants were dying everywhere in Kenya. Bloated faceless carcasses were dotted across Lamu, Tana River, Samburu, and Isiolo districts, across Tsavo, Meru, Kora, and Mt. Elgon National Parks, across Shaba, Boni, and Dodori National Reserves, not only singly, but in groups. Whole families were being slaughtered by poachers with automatic weapons who had moved into the very heart of the parks, where they worked with impunity.

Aerial surveys continued to count the living and the dead. Certain stages of decomposition allow scientists, on an aerial survey, to estimate how long ago an elephant had died. "Fresh" refers to a bloated carcass lying in a pool of blood, topped with the white stains of vulture feces, a hacked-off face, a trunk tossed to the side. "Recent" means less than one year: a shrunken piece of skin draped over a pile of bones surrounded by a bare patch of earth, the "rot patch." "Old" means a pile of white bones gleaming in the sunlight. And "very old" means a pile of gray bones, cracked from years of relentless sun. These are important distinctions when thousands of elephants are being killed every year, as they were in the late 1980s. They can give you clues to patterns in the ivory trade, the methods used by poachers, or the effectiveness of an antipoaching force. But these dead elephants were merely numbers: A dot on a map after an aerial survey may represent ten dead. Imagine 2,259 "old dead" and almost 200 "recent dead" elephants, as were counted in Tsavo National Park in 1988.

Secluded in Amboseli, surrounded by the sounds and smells of unperturbed elephants and engrossed in the exciting new discoveries I was making, I had been able to forget the many problems the elephants faced around the continent: the growing conflict with humans, the poaching for ivory, the compression of elephants into protected areas caused by the poaching and the increasing conflict outside, and the culling that was taking place in parts of southern Africa where it was felt that there were too many elephants. During much of my time in Amboseli, ivory poaching in Kenya had been at a relatively low level. After close to 110,000 individuals had been killed in the five years between 1973 and 1978, there had been a four-year lull in poaching partly as a result of financial input from a major World Bank investment project in the wildlife sector. But then the international demand for ivory had begun to rise again and with it the price of ivory and the killing of elephants. By the time I was in the midst of my communication study in 1987, poaching was raging through much of the country.

In May of that year I attended my first meeting of the African Elephant and Rhino Specialist Group (AERSG) in Nyeri, at the base of the Aberdare mountain range in central Kenya. The AERSG grew out of the Elephant Specialist Group coordinated by Iain in the 1970s and was established in 1982 as part of the Species Survival Commission, a group of scientists who work within the International Union for the Conservation of Nature (IUCN) to protect threatened species of wild plants and animals. Members of the various specialist groups are selected on the basis of their expertise. Cynthia had been a member of the AERSG since 1982, and though I was not yet a member, I had been invited to present the findings of my recent research. The Nyeri meeting was a revelation and was to stand out in my mind in a number of ways. It was my entry into the bitter battle to save the elephants and into elephant conservation politics. Though I had heard all of the names and seen many of the faces of the conference attendees before, it was my first opportunity to discuss elephant issues face to face with the wildlife utilization hard-liners from southern Africa. Like Cynthia, I was already labeled as being "soft on elephants," and I was quickly engaged in debates with scientists who believed that the best way to protect elephants was consumptive utilization: culling "excess" elephants and selling their tusks and hides in order to raise the money needed to pay for their conservation. With the ivory trade raging out of control, this year's meeting promised to be controversial and to focus the hardening of the eastern and southern Africa divide on elephant conservation issues.

It had always surprised me that despite the number of women working in the field of wildlife and conservation biology, university faculties, wildlife department positions, meeting rooms, and conference halls were invariably dominated by men. Of the forty or so elephant participants in the Nyeri meeting room, only a handful were women: As I recall, they were Cynthia, Anne Burrill, who was working with Iain Douglas-Hamilton, Kes Hilllman (who was primarily representing rhinos), and myself. There were no women from southern Africa. I later learned that this was because no women held relevant positions: There wildlife management and research was and still is considered a job for men. I had occasionally mused how odd it was that, despite ten years of elephant experience and a Ph.D. in elephant behavior, I had yet to be invited to become a member of the AERSG. I had assumed that I had not been chosen because the group was primarily concerned with the status of the species and that perhaps my behavioral work had been viewed as unrelated to elephant conservation. But upon seeing the sex ratio in the room, I wondered whether I had simply been overlooked.

As it was, I was there as a guest. Earlier that year I had been invited to participate in the elephant meeting and to present a paper on elephant communication. On the preliminary schedule, I had noticed that I was the only woman making a scientific presentation and the only person asked to do so outside of the regular sessions— in the evening, after dinner. Perhaps the evening slot had been offered to me because I was not a member of the AERSG, or perhaps, again, it had been because my talk was not considered to be directly related to the conservation of elephants. Whatever the reason, at the time I was very sensitive to any form of discrimination against women, and I didn't like the implication that a group of men viewed my work as merely an evening's entertainment. I declined the evening slot but said that I would be happy to present my findings during the regular sessions. The organizers agreed.

The meeting began with the presentation of technical papers, during which I spoke about my discovery with Katy that many elephant vocalizations contained sound at high-pressure levels below the level of human hearing and that elephants were using some of these calls to communicate with others five, perhaps ten, kilometers away. After presenting the purely technical aspect of my talk, I discussed some of the ways my findings related to elephant conservation and management. I suggested that the acute low-frequency hearing of elephants and their ability to communicate over long distances might explain why elephants up to sixty kilometers away from the helicopters used during the Hwange National Park cull, in Zimbabwe, had run to the far corner of the reserve and hidden until the cull was over. I explained to the group that the older males once thought to be "dead wood" were actually the breeding musth males and that the continued slaughter of these individuals to supply the ivory trade with their large tusks could have implications for the survival of the species. Finally I made a plea for the elephants by arguing that the thousands of hours we had spent observing them in Amboseli had shown them to be highly intelligent, sophisticated animals and that they deserved to be treated as something more than just so many kilos of ivory.

Until then the AERSG meetings had revolved around measuring the ecological damage elephants caused, deciding how many could be sustainably harvested and how much ivory or meat they could provide, or counting how many remained dead or alive. Very little consideration had ever been given to elephant behavior and its significance for their conservation. Iain Douglas-Hamilton later told me that during my talk he had noticed several alpha males shift uncomfortably in their seats and that one had acknowledged that what I had said made him wonder whether consumptive utilization was the right way to go. His doubts must not have stayed with him for long, however, for the status quo remained, and any suggestion of a reduction in the consumption of ivory was rejected.

Iain then presented his data on elephant numbers and population trends with his assistant Anne Burrill, a no-nonsense computer whiz who, with Iain, had logged onto the computer huge quantities of data on factors affecting elephant densities across the continent. As I listened to her summarize her Global Information System analyses and watched her pin up map after computer map, Iain's findings were indisputable. Except in a few isolated pockets, the elephant was in devastating decline across the continent.

Later when the meeting broke up into smaller regional groups to revise the estimates of elephant numbers for each country, Iain, David Western, Cynthia, and I worked together on Kenya's population. With populations outside the protected areas already greatly reduced, the poachers had turned to the protected areas, and we estimated that a mere 20,000 elephants remained in the country. When the conference reconvened and the estimates came in from around the room, country by country, with few exceptions, all were revised downward. As we sat recording the demise of the world's largest land mammal and one of its most socially complex species, I became disconcerted by the apparent lack of alarm. We were supposed to be the individuals to whom the world turned for advice about the status of African elephants. Why wasn't anyone doing anything? Why was the prevalent tone one of complacency?

The African elephant had been provided some nominal international protection since 1975 under Appendix II of CITES, the Convention on International Trade in Endangered Species of Wild Fauna and Flora. CITES is a global body charged with controlling international commercial trade in endangered species, and in the late 1980s close to one hundred countries were party to the convention. Each country has a scientific and management authority whose responsibility it is to control imports and exports of different species of plants and animals according to an agreed-upon set of rules. Appendix I species include those deemed by a majority of parties to be threatened with extinction, and under the rules of the convention, member states must not engage in international commercial trade in individuals of these species or their products. Appendix II includes species considered to be threatened, but in whose products the convention allows a limited trade with CITES permits. The effectiveness of CITES, however, depends largely on the amount of law enforcement in both producer and consumer countries. As an indication of CITES' effectiveness, since the African elephant's inclusion on Appendix II, well over half of the population had disappeared.

In 1985 CITES acknowledged that a huge portion of the ivory trade originated from illegally killed elephants, and it adopted an ivory quota system to try to bring the trade under control. The basis of the system was that each country with elephants would declare an annual export quota of ivory, which it would not surpass. Every tusk leaving Africa would require an export permit, and the producer states, in collaboration with the CITES Secretariat, would control the issuance of permits. The idea was that each exporting country's quota of ivory would bear some relationship to the size of its elephant population; in other words, the offtake of ivory would be sustainable. Yet CITES did not actually insist that this be the case, and by 1987 it was clear that many countries were using quotas simply as an ivory export notification system, and there was very little relationship between the natural production of ivory in a country and its export figures. The quota system was, in essence, being used to legitimize the illegal killing of tens of thousands of elephants.

For example, Burundi, with no wild elephants of its own, was one of the most notorious entrepôt countries for illegal ivory. According to Burundi Customs statistics, the country exported 1,305 tons of ivory between 1965 and October 1986. In 1986 CITES "legalized" 89 tons of raw ivory held by Burundi under the Ivory Control System. The only condition laid down by CITES for the legalization was that no more ivory should be smuggled through Burundi. However, despite the agreement with the CITES Secretariat, an additional 110 tons of raw ivory was exported from the country between November 1986 and October 1987. To provide this amount, 11,000 elephants had been illegally killed in Tanzania, Kenya, Zambia, and Zaire.

Until I attended the Nyeri meeting, I had believed that CITES had been formed to protect endangered species from being unsustainably exploited by commercial trade. But the more I listened to the views of its secretariat, and the more I learned about its approach and the various deals it made in the name of "elephant conservation," the more I began to view CITES as a mechanism that was being used to prevent species conservation from endangering international trade in wildlife products.

Despite the evidence, the ivory trade experts at the 1987 Nyeri meeting continued to insist on keeping the elephants on Appendix II of CITES and on giving the ivory quota system a chance to work. While Iain, Cynthia, and I argued that the price of ivory should be kept as low as possible to discourage poaching, Rowen Martin, a senior biologist in the Zimbabwe Wildlife Department, and others from southern Africa stated that, if anything, we should encourage a higher price for ivory, as this would increase the value of elephants to people and encourage their protection. The southern African states believed that they had plenty of elephants because they had managed them properly and that we, in eastern Africa, had very few because we had not. Consequently they wanted to utilize their elephants commercially, and they wanted the price of ivory to remain high. They did not wish to be penalized for what they saw as our incompetence and mistakes.

As I listened to the arguments back and forth, it became clear that there were two very different schools of thought. One was that proper management and so-called rational sustainable utilization of elephants and their products was the answer. In this view, a ban on the ivory trade was unthinkable, because it would mean the end of international commercial trade in ivory, upon which the value of elephants was dependent. The other was that with civil war, political instability, poverty, and corruption rife in Africa, how could anyone reasonably expect that elephant ivory would be used rationally and sustainably? This contingent believed that there was no future for the elephant as long as the price of and demand for ivory remained high and, therefore, the buying, wearing and selling of ivory should be discouraged. The two sides were on an inevitable collision course.

Cynthia, who had grown angry with the direction the discussions were taking, commented that not all of us in the room agreed that elephants should be utilized consumptively. Steve Edwards, an Executive Officer of the Species Survival Commission, replied that not to utilize elephants would be contrary to IUCN's World Conservation Strategy on sustainable utilization, which regarded man and his needs as the focus for all conservation endeavors. There was no discussion of whether their so-called sustainable utilization theories were in fact sustainable in practice.

Cynthia and I were disturbed and frustrated by the meeting, and we joined the Douglas-Hamiltons at their home outside Nairobi to plan a strategy to campaign for the end of the ivory trade. To ban the trade completely, the parties to CITES would have to be persuaded that the Africa elephant was in danger of becoming extinct. We knew that the ivory trade issue was highly contentious and that winning an Appendix I listing for elephants at the next convention would require a major public awareness campaign. We also realized that a ban without a corresponding reduction in demand would only force the trade underground. We would have to reach the general public and persuade people around the globe not to buy, sell, or wear ivory.

Cynthia and I believed that the most effective way to reach the public was through the media. If we could get enough media interest in the Amboseli elephants, we could send out a powerful message against the ivory trade. Amboseli's elephants are so docile that visitors can approach them to within a few meters, and the park's relatively easy access to Nairobi made it an obvious choice for film crews and journalists. As the ivory trade issue gained momentum, Cynthia and I put our research aside whenever a crew requested assistance in locating elephants, interpreting their behavior, or obtaining an interview. Between 1988 and 1990 we hosted literally scores of film crews from all over the globe, and I think it would be fair to say that up to 75 percent of the footage of elephants obtained during the late 1980s was filmed in Amboseli. On one impossible day in early 1989 five separate crews arrived, each wanting to film the elephants. Thankfully, most of these visits were short and we were able to continue with our research.

During September 1987 Cynthia and I visited the United States. We met in Washington, D.C., in the offices of the African Wildlife Foundation. From the inception of the Amboseli Elephant Project in 1972, the African Wildlife Foundation had provided a "home base" for the project and financial assistance for its operations. Cynthia and I met with the director, Paul Schindler, and explained why we believed that an ivory trade ban was the best way to protect the African elephant. We then proposed that the African Wildlife Foundation work with us in a public awareness campaign. After we received support from the foundation, we set out to discuss our ideas with the many international conservation organizations that are based in Washington.

The first organization we visited was the World Wildlife Fund (WWF). Buff Bohlen, a veteran conservationist and later head of President Bush's Environmental Protection Agency, agreed to see us. We explained why we believed that an ivory ban was the best solution under the circumstances and why we were confident that it would work. Buff replied that we just didn't understand and insisted that the ivory issue required a "rational" approach. He asked us to listen to the WWF's ivory trade expert, Jorgen Thomsen, on why the trade was good for elephants. Jorgen was an officer of TRAFFIC, a wildlife trade monitoring organization backed by WWF and IUCN.

Tall and dark with deep blue eyes, Jorgen looked at us intently and, in his deep, slow voice, tried to speak convincingly in favor of the ivory trade. There was nothing new in what he said, and there was nothing that would stop the slaughter of the elephants. We had heard his hollow words before; they were being repeated by other conservationists in offices all over the world. Once again we were also told that our sentiments about elephants being complex, highly social, intelligent animals was an emotional Western notion, and one foreign to Africans who, apparently, only understood money.

Cynthia and I left the plush WWF offices disappointed. While this was the organization that mass-mailed highly emotional pleas for support for their efforts to save rain forests, pandas, whales, and elephants to almost every household in America, its policies didn't jibe with its message. In trying to please every one of its constituents, from little old ladies concerned about cruelty to animals to those who argued for the culling of elephants for "rational sustainable development," it was forced, in effect, to sit on the fence. These were the individuals who would decide the fate of the African elephant, and they knew next to nothing about Africa or about elephants. They were, it seemed to me in my anger, like so many others: the gray men in the corridors of conservation power.

CHAPTER ⇒ 24

Dire Predictions

ARE WE NO LONGER CAPABLE OF RESPECTING NATURE, OR DEFENDING A
LIVING BEAUTY THAT HAS NO EARNING POWER, NO UTILITY, NO OBJECT
EXCEPT TO LET ITSELF BE SEEN FROM TIME TO TIME? LIBERTY, TOO, IS A
NATURAL SPLENDOR ON ITS WAY TO BECOMING EXTINCT.
—Romain Gary, *The Roots of Heaven,* 1958

I RETURNED TO AMBOSELI late in the year to continue with my
research. In February 1988 Iain Douglas-Hamilton invited me to
join an aerial count of the Tsavo elephants that he was organizing.
It took nine single-engine aircrafts five days of flying to cover the
22,000-square-kilometer Tsavo National Park and adjoining elephant habitat. Flying with Iain is memorable by any standards, but
counting elephants as his front-seat observer is better experienced
under the effects of Dramamine. Iain's takeoff was accomplished at
full throttle, over the shortest distance possible and with the aircraft
protesting loudly. Once aloft, whenever we circled tightly over a
group of elephants, the stall warning buzzed continuously. Landing
was especially dramatic: We flew the length of the runway just
meters off the ground and then banked sharply, turning, it seemed,
almost upside-down, and then landed into the wind. This survey
was the first major elephant aerial total count I had been involved
in and, being prone to airsickness and preferring to watch elephants
from the ground and up close, I have tried to avoid them since
then.

During our time at Tsavo I stayed with the Douglas-Hamiltons at the house of wildlife filmmaker Simon Trevor, just inside the gates of the national park. It was there that I met a strong-willed and dynamic young woman with sun-bleached hair and captivating blue eyes named Barbara Tyack. She had started her career in wildlife by studying baboons in the Mauritanian desert, though she was determined one day to become a camerawoman. After meeting Simon she first convinced him that she was a good cook and then talked herself into being accepted as his camera assistant. Two years later she was making her first film. Barbara was a woman after my own spirit, and we established a close and lasting friendship.

As Iain's front-seat observer, it was my responsibility not only to keep careful records of the number and location of both the live and dead elephants we counted but also to plot the groups and the bones on a map when we returned at the end of the day. As I sat on Simon's verandah methodically mapping the data, I saw an ominous pattern emerging: A significant number of the dead elephants were lying within a hundred meters of the road. This pattern could mean only one thing: Elephants were being shot from vehicles.

I remembered that Barbara had mentioned that the elephants were very frightened of the Toyota Land Cruisers used by the park's personnel. It was rumored, too, that elements of the WCMD were corrupt. Then it came to me in a flash accompanied by a wave of nausea, and I wondered how I hadn't realized it before: Some of Tsavo's rangers were actually using their own vehicles to go out and shoot elephants. I called Iain to come and look at my map. Without revealing my suspicions, I asked: "Isn't it interesting that most of the carcasses are so close to the road? Why do you think there is such a nonrandom distribution?" Iain called Patrick Hamilton, a previous warden of Tsavo and dedicated conservationist, over to examine the map. As I expected, they looked at each other in grim silence. My fears proved to be justified; eventually it was revealed that the WCMD employed a number of corrupt rangers who not only colluded with the poachers, providing vehicles to transport their ivory hauls and taking a cut of the profits, but in many cases they were apparently killing elephants themselves from their own vehicles, vehicles that had been provided to protect elephants.

At the end of the count we had tallied a minimum of 5,363 live elephants and 2,421 dead ones. We calculated that somewhere between 5,000 and 7,000 animals had probably been killed in Tsavo over the previous eight years—an average of two each day. Iain's subsequent report, coauthored with Perez Olindo, then director of the WCMD, and Patrick Hamilton, and entitled "The 1988 Tsavo Elephant Count," was distributed widely. While it created a stir in all of the conservation organizations, for many weeks the Ministry of Tourism and Wildlife of the Kenya government was silent.

Through 1988 the poaching situation continued to worsen, and another three international elephant conservation committees were formed to address the crisis: the African Elephant Conservation Coordinating Group (AECCG), the Ivory Trade Review Group (ITRG), and African Elephant Working Group (AEWG) of CITES. It wasn't always clear to me what the specific mandates and goals of each of these "alphabet" groups were, nor why so many groups were necessary, but one thing was apparent: I wasn't invited to any of their meetings. The AECCG was composed of elephant experts and representatives of donor organizations who began to meet regularly to put together an African Elephant Conservation Action Plan. As Cynthia and I had studied elephants for well over a decade, we had ideas to contribute, and we were disappointed not to be invited to participate. Meeting after meeting went by with one excuse after another being given as to why we were not included. I became more and more frustrated and angry that policy decisions were being made about the future of the African elephant, and we were given no voice.

In July 1988, almost a year since our visit to the WWF, Jorgen Thomsen came to Kenya for a meeting of the AECCG. Afterward he and two others from the meeting visited Amboseli, and I took them out to see the elephants. Still annoyed at my apparent exclusion from the elephant meetings and still suffering from the emotional scars of "the incident," I tended to view the world and its male inhabitants as hostile and most of their actions as simply an expression of male domination. I told Jorgen that I thought the decision to not invite us to participate was sexist. Jorgen mentioned my sentiments to David Western, who was by then both chairman of the African Elephant and Rhino Specialist Group and coordinator of the AECCG. David was offended by my accusation and came to explain his decision to me. As I had known him for so many years, he told me, I should have known that he, of all people, was not sexist. To be fair, David was right, and I was probably overreacting. The reason, he explained, that I hadn't been invited to attend the meetings was because my work was "not relevant to elephant conservation."

In one's life, a few statements cause such deep anger that they gain a hold in one's mind, playing themselves over and over again. Often such anger goads one into action, and this was one such occasion. As I sat in the car at the Amboseli airstrip listening to his words, I thought to myself, *I will show you that my work is directly relevant to elephant conservation.*

August 1988 was a grim month in a grim year. Although seven months had passed since the Tsavo elephant count had brought the poaching situation into clear focus, the slaughter continued unabated. I felt helpless: I could not fight the poachers in the bush nor the corrupt elements in the WCMD; nor, it seemed, could I have a voice in deciding the fate of the elephant. Tucked away in Amboseli, I now found it almost impossible to concentrate on my communication study. The more I learned about these fascinating, intelligent animals, the faster they died. I couldn't sit back and watch any longer, I had to do something.

The East African Wildlife Society had scheduled a wildlife conference in Tsavo for the second week of September, and Keith Lindsay had written to ask whether I would present a paper on elephant feeding ecology on his behalf. I had agreed. Now I sat staring at his slides of content Amboseli elephants feeding on acacias and graphs of vegetation biomass and elephant numbers. As I held up one slide after another to the light I realized that I would find it difficult to present his paper in Tsavo in 1988, when elephants were being slaughtered at a rate of two to three every day. I had an idea: I would use the opportunity instead to speak out about the effect that ivory poaching was having on the elephants' social and reproductive behavior and the consequences for the survival of the species.

I quickly pulled out the long-term records from Amboseli and worked late into the night analyzing data, drawing graphs, and selecting my slides. From Cynthia's demographic records I calculated that, under natural conditions, no orphaned calves under two years old had ever survived their mothers' death. Of those between two and five years old, only 30 percent had survived the two years following their mothers' death; and of those between five and ten years of age, only 50 percent had survived. Pili's published work had showed that calf survival was significantly lower in small families with fewer young "aunties" to assist in the care of calves, and from Cynthia's work I knew that the death of a matriarch often caused the disintegration of family and bond groups, thus affecting overall survival. From my own research on the strong preferences of estrous females for musth males and the inexperience of young males, I predicted that the loss of older males would reduce the reproductive rate and a population's potential for recovery from poaching. I didn't prepare Keith's paper; I didn't want to have any excuse for not speaking out about the poaching. I hoped that he would understand.

As it happened, the timing of the talk was perfect. A huge public row about the elephant slaughter had just erupted in the press between Richard Leakey, world-renowned paleontologist and director of Kenya's National Museums, and George Muhoho, the minister of Tourism and Wildlife. Leakey had entered the fray, attacking the ministry for its complacency over the poaching of elephants, condemning it as "economic sabotage." Responding to an earlier press conference at which the minister had expressed "regret" at the killings, Leakey said that the poaching was a national crisis, "not merely something to be regretted." He claimed that high-ranking people were known to be involved in the poaching and that political sensitivity was thus the ministry's biggest handicap in dealing with it. The minister had in turn advised Leakey with his "cheeky white mentality" to stick to his field of prehistory. I had never met Richard Leakey personally, but I admired his courage. I couldn't think of anyone else I knew who would risk his job and perhaps his life to attack the government openly on the ivory issue. The fight between Leakey and the minister made the elephant crisis major news overnight, and articles appeared in the local papers day after day. On my way to the Tsavo conference I added my voice to the debate by meeting briefly with an editor of the *Weekly Review,* a magazine on topical Kenyan issues, and handing her a copy of my intended presentation.

The cover of the September 9, 1988 issue of the *Weekly Review,* featured a gruesome photograph I had taken in Amboseli of a young male with his face hacked off by poachers, and the accompanying story contained the statistics I had given during my Tsavo talk. By then President Moi himself had spoken out against the poaching of elephants and had assigned a wing of the police, the crack General Service Unit (GSU), to assist game rangers to fight the poachers. Meanwhile Leakey continued to attack the ministry; he said that although it had in its possession a document that named senior officials involved in the poaching and trade, it had taken no action against them. After a group of tourists was attacked by armed bandits, the president issued the now-infamous "shoot-to-kill" policy on poachers. The battle for Tsavo had begun.

Within a week four men had been caught with nineteen tusks, but poachers also had shot nine GSU men. Day after day the struggle for Tsavo raged on, with the GSU patrolling the park in helicopter gunships. In Nairobi the police had begun to question wildlife officers over their involvement in poaching, and by the end of the month seventy individuals were dismissed. Finally the elephants had been granted reprieve: Poachers were on the run, and the government was getting serious about dealing with WCMD abuses.

There was progress on another front, too: Jorgen, once so apparently in favor of the ivory trade, had written to thank me for introducing him to "all the pachyderms" and to admit that "feeling the harmony of the Amboseli elephants made a few things change place at some deep level inside." He said that he would be back in Kenya in late October to participate in the first meeting of the CITES African Elephant Working Group and the second meeting of the Ivory Trade Review Group, and he expressed his desire to stay a few extra days so that he could spend more time watching elephants. CITES had commissioned the AEWG to examine the evidence that the ivory trade was seriously affecting elephant survival, and the group included representatives from both producer and consumer member states as well as a number of elephant and ivory trade experts. The ITRG was an independent body coordinated by Steve Cobb, an old friend of Cynthia's with a dry sense of humor. The group was made up of elephant biologists, ecologists, population biologists, ivory trade specialists, and economists who were drawn together to review the impact of the ivory trade on elephants in preparation for the seventh CITES Conference due to be held in October 1989.

In late October I met Jorgen in the Nairobi Offices of the WWF, and we set off for Amboseli for two days of elephant watching between the elephant conservation meetings that he was in Kenya to attend. Cynthia and I had agreed that it was worth taking the time to introduce Jorgen to the elephants and to try to convince him that a ban might be a better way to go. To spend an afternoon in the close company of elephants is awe-inspiring, but to watch elephants in the company of someone who knows each individual by name and can describe the meaning of each interaction, posture, and vocalization can, I have been told, be a very powerful and moving experience. It takes a very callous person to witness these animals and not be led to some fundamental questions about humanity and our place in the world. The elephants gave Jorgen one of their better shows and, as we sat watching babies playing and young allomothers, or "aunties," ensuring they didn't get into trouble, I sensed that the scene was affecting him deeply.

Jorgen had to leave Amboseli the next day to attend the meeting of the African Elephant Working Group. As Cynthia was returning to Nairobi, Jorgen traveled with her while I remained behind in camp. Later that day I received a message from Cynthia saying that I should leave Amboseli before sunrise the next day in order to be at the AEWG meeting, which started at 9:00 A.M. Jorgen had thoughtfully asked the chairman of the AEWG, then Perez Olindo, whether I could attend as an observer, to which he had agreed.

At the meeting data were presented to demonstrate that, at the current rate of killing, or "offtake," eastern African elephant populations would reach "commercial extinction" by 1995 and those of the continent as a whole, by the year 2015. Despite the dire predictions expressed, no one suggested any more radical interventions to protect the elephants. For CITES to list a species on Appendix I was to ask a trading convention to ban trade in one of its products; Appendix I placement was, in effect, an admittance of failure. Twice during the meeting Jorgen warned that at the next meeting of the parties to CITES, only a year away, he thought it likely that some country would propose that the African elephant be moved to Appendix I, thus banning the trade in ivory, and he asked the various country representatives to consider how they would respond to such a proposal. Twice his comment was ignored. Apparently, from the reaction at the meeting, a ban on the trade in ivory was still unthinkable. But Jorgen's mind *had* been turned, and that evening I met with him and other sympathetic parties to discuss the possibility of putting together a formal proposal to have the elephant moved from Appendix II to Appendix I of the convention.

On November 18, 1988, Jorgen wrote from Washington, D.C., to say that he had finally come to realize that the African elephant would be gone very soon in most areas. "The industry," he wrote, "would call it commercially extinct. A frightening concept that will be applied to most of the large terrestrial mammals in our lifetime." Jorgen admitted that he had come to believe that the only way to proceed was to propose the transfer of the African elephant from Appendix II of CITES to Appendix I, and he suggested that a draft proposal be written in Nairobi with Iain's assistance no later than February 1989. He ended his letter by asking whether I would be willing to help. I was thrilled to be given an opportunity to participate and agreed to his proposal without a moment's hesitation.

Our activities, he said, would have to remain a secret, because the majority at the WWF still believed that an ivory-trade ban was not necessarily in the best interest of Africa's elephants. While I had nothing to lose by becoming involved, as a representative for the WWF Jorgen was overstepping his bounds, particularly as he had not received official permission or even informed the organization of his plan.

In late 1988 I was in the midst of my elephant vocal communication study, but I wanted to gather some facts about the effect of poaching on the elephants' social and reproductive patterns. I knew from the number of elephants being killed, the number of orphans being reported, and the rapidly declining average weight of tusks being traded (which showed that younger elephants and a higher proportion of females were being killed) that the situation throughout much of eastern Africa was dire, but I wanted to have facts and figures from areas under heavy poaching, not just theory based on Amboseli data. For that I needed to go to where the poachers were operating; I needed to see the elephants for myself, to age and sex them, and to watch their behavior. I wrote to the National Geographic Society, which was supporting my communication research, to ask if I could interrupt my study for several months, and it agreed.

Cynthia was not happy with my plan. For several months she and I had talked vaguely about doing a series of surveys together, but in late 1988 I told her, perhaps rather abruptly, that I wanted to do them on my own. I tried to explain my reasoning to her, but apparently I did not succeed for she became very upset with me. For a long time I had felt a growing concern over my future. What was I going to do? I had spent my entire adult life studying elephants in one small national park in Kenya. Elephants were my passion, and I had never held what most people would consider a "proper job." At Princeton I had been encouraged to apply for academic positions, but I had declined, knowing that my future was in Africa. But now it was becoming very clear that I couldn't stay on in Amboseli forever, living from one subsistence grant to another. I knew, too, that I would never run the Amboseli Elephant Project. Cynthia had started the project and the elephants were her life. She would continue as director for as long as she could. If I was to obtain one of the few positions in a Nairobi-based conservation organization, I knew that I would need to gain more experience and broaden my background. In short, I needed to prove myself in an area other than pure research. If Cynthia and I carried out the surveys jointly, they would, from my perspective, be credited to her and I would be no closer to sorting out my future. From Cynthia's perspective I was just being incredibly selfish, and she told me so in no uncertain terms. She also told me that my decision would alter our friendship forever. It was a classic case of a student-mentor rift. But our discord ran deeper than that because since the time of my father's death and my family's departure from Kenya, Cynthia had always acted as a parent to me, giving me wise counsel on my relationships and always offering sensible advice in dealing with my day-to-day problems. Cynthia had always been matriarch of the project, the camp, and, in many ways, my life, and now I was breaking away. The intensity of her reaction was devastating to me, but I could not turn back. I proceeded, knowing that she felt that I had betrayed her and that I had lost her wholehearted trust and support perhaps forever.

The African Wildlife Foundation, which by now favored an ivory trade ban, agreed to sponsor my study, which was to involve a survey of Tsavo National Park where poaching had caused a decline from some 45,000 elephants in the late 1960s to some 5,000 by 1988; Mikumi National Park, the northern extension of the vast Selous Game Reserve in Tanzania where, over the entire ecosystem, elephants had been reduced from 110,000 to 30,000; and Queen Elizabeth National Park, in Uganda, where Idi Amin's troops had decimated the elephant population in the late 1970s. My intention was to compare the age structure and sex ratio of each of these populations with the relatively undisturbed Amboseli elephants.

I planned to carry out the Tsavo survey in January, work with Jorgen on the Appendix I proposal in February, and then survey the Mikumi and Queen Elizabeth elephants in March and April. Permission to undertake the surveys in Tanzania and Uganda was granted by the appropriate authorities. Kenya was more problematic. I had research clearance for a study of elephant communication in Amboseli, but I suspected that, in the current climate of mistrust, obtaining permission for the Tsavo study would be unlikely. The deteriorating security situation in Tsavo was affecting tourism, Kenya's number-one source of foreign exchange. Bandits armed with automatic weapons had the run of the park; elephants were being slaughtered, and tourists had been shot at, some killed. The government was particularly sensitive about Tsavo because of the bad press it had been receiving both internationally and at home, and journalists and film crews were not being allowed to cover the park. I broached the subject with Perez Olindo, then still director of WCMD, who had the authority to grant permission for short-term surveys such as the one I was planning. He agreed that I could go ahead, but he would not give me anything in writing. I would be on my own.

CHAPTER ➤ 25

A Poacher's Story

NILIANZA KUSIKIA MBAYA, KUTETEMEKA, DUNIA IKAWA GISA.
I STARTED TO FEEL BAD, TO TREMBLE, THE WORLD BECAME DARK.
—Arrghas Ore, explaining why he had
abandoned poaching elephants, December 1988

I SPENT CHRISTMAS OF 1988 at the coast with Barbara and the Douglas-Hamilton family. Despite the misunderstanding with Cynthia I was in particularly high spirits, for concurrently with my increasing involvement in saving Africa's elephants, I had fallen deeply in love. I frequently caught Oria watching me, a twinkle in her eye and a warm smile on her lips. Oria understood the wild, romantic, impulsive side of my nature, and as she took me for long walks on the beach, her arm around my waist, she would say, "Tell Mama Oria." The man I had met was tall, handsome, kind, and understanding, and after all of my solitude and anger, I couldn't believe my good fortune. From a young age I had wanted children, and I had dreamed that I would one day marry a man with whom I could share my unusual African existence. Now it seemed that my dream would be fulfilled.

Then during that Christmas holiday with the Douglas-Hamiltons I met an extraordinary man named Arrghas Ore. A mutual friend of ours had introduced us, because he thought Iain and I might be interested in what Arrghas had to say. An Orma from the north-eastern border of Tsavo, Arrghas was tall and handsome in a rugged way. He was a man who thought nothing of walking three hundred kilometers from the Kenya-Somali border through the dense bush of Lamu and Tana River districts to Tsavo. At the border town of Kolbio he could buy a Russian made Kalashnakov, or AK47 automatic rifle, from deserters of the Ogaden war for a mere fifteen dollars. Arrghas was a nomad, and killing elephants was how he made a living.

Arrghas told us that he had started poaching elephants in Tsavo during the late 1970s as a young boy; he had abandoned the national park only four years later, when a dedicated warden, Joe Kioko, began daily aerial patrols. But when, during the mid-1980s, the WCMD and the antipoaching unit started to collude with the poachers, he returned. During the intervening years he had plundered the elephants of the remote Boni Forest, tucked up against the Somali border, until there were none left. In those days killing elephants brought good money. *"Ilikua na bei,"* he told us. You could buy trucks, build houses with the money from ivory.

Up until February 1988, the price of ivory on the local market had been around 25 dollars per kilogram, and there was a good market for it. Each elephant produced an average of ten kilograms, and with automatic weapons poachers were able to gun down entire families quickly. The collusion of some of the WCMD staff made their job easier. In the late 1980s, Arrghas alleged that a warden provided transport for poachers, ensured that the security forces were sent out in the opposite direction, and arranged ivory deals in exchange for a cut of the profits. Members of his staff sometimes shot elephants from the roadside in their own vehicles.

As Arrghas recounted his tale, I thought back to what I had observed during the 1988 aerial count of the Tsavo elephants: the terrified elephants running with their tails in the air as soon as they heard the distant hum of the WCMD's Toyota diesel engines; the maps of live and dead elephant distributions revealing the high proportion of elephant carcasses lying next to the roads. The look that had passed between Patrick Hamilton, one-time warden of Tsavo, and Iain Douglas-Hamilton had made my heart skip a beat. Elephants had indeed been killed by the very people charged with their protection.

Arrghas went on to say that after the Tsavo elephant count, the corrupt members of the WCMD began to get nervous and were less willing to cooperate. It became more difficult for the poachers to get ivory out of the park and to locate buyers. The dealers, too, were afraid of being caught and began to grow more cautious.

By April 1988, Arrghas told us, the price of ivory had fallen to 10 dollars per kilogram. He related how he and his gang sold their ivory, went to Mombasa, bought a new set of clothes, perhaps a pair of sunglasses, and then the money was gone. He demonstrated how quickly the money ran out in the African way: a burst of air as he flicked his index finger quickly over his lips. Once the money was spent, they returned to Tsavo to kill more elephants.

Arrghas, Iain, and I spent an entire morning talking. Arrghas described the life of an ivory poacher and told me a story about the last elephant he had killed and why, in August 1988, he had turned himself in to the authorities.

In the bush, members of a poaching gang left intricate messages for one another: shapes drawn in the red Tsavo dust with numerous figures scribbled inside that looked like a foreign language or the wrinkle pattern of an elephant's footprint. If the lines added up to nine, then the coast was clear; if they added up to ten, then the security forces were nearby and it was not safe to move. To confuse the forces, the gang walked back and forth, off in one direction, all the way back in another, and round and round in circles. When they finally left they traveled on the grass, leaving no tracks at all.

At the end of a raid, they carried the ivory out of the park and buried it, carefully camouflaging their cache with grass and old elephant dung. They situated their caches between three trees, each one marked with the cut of a machete. One gang member was designated to look for a buyer, who would be driven straight to the site. Then the tusks were uncovered and weighed on the spot.

The ivory went first by car and then by boat or airplane to Somalia, Tanzania, or Burundi.

Sometime in mid-1988 Arrghas shot a female elephant with a one-day-old baby. As the mother fell to the ground she landed on top of her calf, killing it, too. Arrghas said that after killing the mother, he was unable to hack the tusks from her skull. He sat under the shade of a tree, away from the dead elephants. *"Nilianza kusikia mbaya, kutetemeka, dunia ikawa gisa,"* he said: I began to feel bad, to tremble, the world became dark. Not to lose face, he told his companions that he was unwell.

Two days later a man came with a message for him from home: His wife, his two children, and his brother's son had all drowned in the Tana River when the boat they were in capsized and rolled over on them. They had died on the same day as the mother elephant and her baby. He told me, *"Ndovu ni kitu mbaya sana."* Elephants are bad, he said, they have much power, they are one with God.

It was then that Arrghas decided to stop killing elephants. He tried to persuade his brother, Golo Ore, to stop, too, but Golo responded by beating Arrghas on the chest with the butt end of his G3 rifle. When I met him five months later he still bore the scars from that attack. After that incident Arrghas left the gang, walked through Tsavo, and turned himself into the police at the town of Manyani along the park boundary.

Arrghas told me that on September 9, 1988, an orphaned baby elephant walked into the mosque at MacKinnon Road on the boundary of Tsavo and stood by a sheik's coffin. On that same day six poachers were shot dead by the security forces. One of them was his brother, Golo. The next day three more poachers were killed, and the following day another three. The elephants were finally getting their own back, he said. *"Jangili wanamalizwa."* The poachers are being finished.

These were not the only stories that made Arrghas believe that elephants had a connection with God. He spoke of trucks bought from ivory profits that crashed, and how the children who lived in houses built from ivory—"*wanakula mavi*"—became mad (literally, they eat excrement). Arrghas had once watched an elephant trying to cover a dead companion with a branch of a tree. When the wind changed direction the elephant caught his scent, and so Arrghas moved some distance away. When he returned he found the dead elephant completely covered with vegetation. Though the elephant's tusks were still intact, Arrghas was afraid to remove them because he felt that God had intervened.

The others in his gang knew that Arrghas had turned himself in, that he had told the authorities how to catch them. They knew that he had been the one who had organized the ambush in which Golo had been shot dead. After our meeting I never saw Arrghas again. A poacher turned informer is a wanted man. Over the years I have often thought of Arrghas and wondered whether he was killed by the remnants of his own gang or whether he returned to the bush to prey upon elephants once again.

CHAPTER ⇒ 26

Tsavo Survival

IT WAS NO LONGER A MATTER OF DEDUCTION OR OF READING DANGER
SIGNALS; IF THERE WAS DANGER IN THE AIR, YOU COULD SMELL IT,
JUST AS YOU COULD READ THE MINDS AND MOODS OF ANIMALS.
—Vivienne de Watteville, *Speak to the Earth*, 1935

I LEFT NAIROBI FOR Tsavo on January 10, 1989, with strict instructions from Perez Olindo to avoid working in areas where the
Government Service Unit (GSU) was carrying out its antipoaching
operations. The GSU had been deployed in Tsavo with authority
to shoot poachers on sight since the previous September, and as
enough tourists already had been killed to affect tourism seriously,
the last thing Perez needed was to have a foreign woman researcher
caught in a GSU shootout with poachers.

I estimated that the Tsavo survey would take just over three
weeks, approximately two weeks for Tsavo East and another week
or so for Tsavo West. I would spend the first ten days staying at
Simon Trevor's house near Voi. Although Simon was away, I
would have the company of my new friend, Barbara Tyack. Since
I would be working from friends' houses or lodges for the duration
of the survey, I didn't need much in the way of safari equipment,
other than for emergencies. I carried extra food and water and a
blanket, and for the survey itself I had a spotting scope, binoculars,

a Dictaphone, a camera, film, data sheets, pens and pencils, and a calculator.

Simon had left a message with Barbara for me: I was to make sure that I told someone in which direction I was headed each day. If I got stuck, if my car broke down, I was not, under any circumstances, to attempt to walk. Tsavo was too vast, the sun too hot, the lions too aggressive, and my chances of survival too slim. I would be better off waiting with my car until someone found me. The following morning I set off on my survey full of the spirit of adventure. I was going to try to drive every road south of the Galana River, an area of some five thousand square kilometers, in ten days, and age and sex every elephant I saw. It was not going to be easy, for the bush was thick and the elephants were terrified.

I left early every morning working my way down each road, in some cases stopping to rebuild a section that had washed away so that I could go forward. By the end of my stay with Barbara I had aged and sexed 503 elephants. The age structure and the sex ratio of the Tsavo East elephants was not good: Very few elephants over forty years old remained alive, and of the population over fifteen years old, 82 percent were females and only 18 percent were males. Males are always the first to be killed since their tusks are so much larger than are those of females. For example, by the age of fifty-five a female's average tusk weight is seven kilograms; tusks of a male of the same age weigh forty-nine kilograms. A male in his late teens already has tusks larger than those of most adult females. What was most alarming about the elephants was the fact that the females and families had been so affected. According to my data, only 37 percent of the families I saw seemed intact, with a matriarch over thirty years old, while 28 percent of the families were obviously missing adult females (as deduced by the numbers and ages of calves relative to adults) and 3 percent of the groups were composed entirely of orphaned calves. But the area south of the Galana River was not nearly as bad as it was north of the Galana, where poachers had all but decimated the elephant population. For security reasons I was not allowed to survey north of the Galana, but

Joe Kioko, who had been sent back to Tsavo to battle the poachers, flew me over the area one morning. From the air we saw small groups with a high proportion of young; most of the adults were gone. We found the largest concentration of elephants on top of the Yatta Plateau, and Joe gave me permission to cross the Galana and spend one day looking for elephants there.

From the air we had seen the groups of elephants easily, despite the plateau's dense vegetation, but once on the ground, my view was blocked by a tangle of branches and thorns. I was assigned an armed ranger for the morning, and as we stopped every kilometer or so I could hear and smell the elephants nearby. I climbed several trees, but even from the vantage point of five meters up I couldn't make out any of the large gray shapes I was searching for. Just as we were about to turn back, I spotted a familiar object on the ground in among the fresh piles of elephant turds: the amniotic sac of an elephant born early that morning. In Africa this is a sign of good fortune. I carried it away with me to the next segment of my survey: Tsavo West National Park.

I was surprised by what I found in Tsavo West. I had been led to believe that the eastern part of Tsavo had suffered more poaching than the west, but as the days passed, I began to wonder whether those claims had been made because Tsavo East had always been the focus of greater interest, whether in the days of "too many" elephants or now in the days of too few. In Tsavo West I aged and sexed an additional 338 elephants, ending up in the southernmost end of the park along the Tanzanian border at Lake Jipe.

Most of the elephants ran as soon as they caught my scent, so I had to complete my work before they were aware of my presence. But one memorable group of forty strong charged me repeatedly. In order to age them I needed to have the elephants either looking at me or moving perpendicular to me. The matriarch of this group was extremely aggressive, and if I moved toward the group or sat still she led the entire group to attack in a solid wall of elephants. Even though I trust most elephants, this was not a challenge to take lightly. With the car engine running, my feet resting on the clutch and the accelerator, I spoke into my Dictaphone until the last second: "female class 3B, 0B, 0A, male 1A, female class 4 . . ." Then I roared off in the opposite direction. I repeated this sequence over and over again until I had aged the entire group.

As I summarized my data at the end of each day, I was surprised to find that by each of my criteria, the Tsavo West elephants appeared to have been more affected by poaching than those of Tsavo East. Of the family groups that I observed, only 29 percent were intact with a matriarch over 30 years old; 25 percent were intact with a younger matriarch; 32 percent were obviously missing adult females; and 14 percent were mostly orphaned calves. The males made up only 16 percent of the adult population, and among breeding-age adults (females over ten and males over twenty-five years old), 97 percent were female and only 3 percent were male.

On January 25 I decided I had collected enough data and that I would head back to Voi and Simon's house by way of Kasigau, on the southeastern boundary of Tsavo West and the last place I had left to cover. I asked the assistant manager of Lake Jipe Lodge to radio Simon that I would be staying with him that night, arriving from Lake Jipe via Kasigau. Later Patrick Hamilton told me that this was one of the most dangerous areas of Tsavo for me to be in: It was full of armed Somali poacher-bandits, and no one else.

Several days earlier, near Maktau, north of where I was now headed, I had met a section of GSU eating lunch under the shade of a broad *Acacia tortilis*. The sergeant offered me a tin of pineapple juice, and so I stopped to chat with them. He thought I was completely mad to be driving around in the area alone and said that he would not go anywhere in Tsavo without a gun. I had laughed then and told them that I wasn't worried, as I wasn't afraid of wild animals and I was used to being alone in the bush. In truth I was concerned about meeting poachers, but I felt that the presence of a ranger or the possession of a gun would only make me more of a target.

I left Lake Jipe at 9:00 A.M. and headed down a small track toward Kasigau. After driving for almost an hour, I came to the spot where I had left the elephants the night before. I could still see the tire tracks where I had driven off the road in pursuit of them and the tree under which I had found a complete skull of a female with her two tusks still intact. Although I had planned to take the tusks to park headquarters, as I had decided to take the road to Kasigau, I left them with the rangers at the Lake Jipe outpost.

Now I had ivory on my mind again. In Amboseli, I often used to get premonitions before I found a tusk. While I cannot explain these feelings scientifically, when I thought I would find ivory, I usually did. I now had that same feeling. I drove on several kilometers farther until I saw a large shiny white object lying in the long grass by a water hole. It had to be another elephant skull. I stopped the car, climbed on top of the roof, and scanned the area for ivory. I couldn't see any, but the skull was a hundred meters away on the other side of the water hole and any tusks might have been hidden in the long grass. I decided to drive closer and have a look around. I got back in the car and turned the key, and nothing but a loud *tick-tick-tick* resounded in the silence of the Tsavo bush. My heart sank. I checked the odometer: forty kilometers from Lake Jipe. I tried again; *tick-tick-tick*. I tightened the battery connection, but it made no difference. I checked the ignition switch, the coil, the starter relay: nothing.

I worked on the car until noon. It was late January, the hottest time of year, and the sun beat down unforgivingly. I tied my *khanga* between the side of the car and two bushes to give myself a bit of shade. I was already beginning to feel insecure and very alone. The track I had come along looked as if it hadn't had any traffic for months. I pulled the recent love letters out of my bag and reread them, and then to pass the time and to keep my mind occupied, I analyzed data through the heat of the day, turning my attention to the car again during the cooler hours of the late afternoon. After an hour or so I gave up and walked over to the skull to see whether I could find more tusks. I discovered one, which, by its slender curve, had belonged to a female. I carried it back to the car with mounting apprehension. Finding another tusk lying right next to a skull reinforced my belief that no one had been along this road for months, perhaps years, and that no one would ever come upon me here, not even poachers.

Dusk crept in. I ate more of my food and drank the rest of my tea. The food was hard to stomach. I don't like croissants at the best of times, and these were doughy. As it started to get dark, I prepared for the night. My original plan in the case of an emergency had been to sleep on the roof of the car, and I was prepared with a thin foam mattress and a blanket. I climbed onto the roof and pulled the blanket around me, aware of a rising unease. As I lay staring up at the stars and listening to the night sounds, nocturnal birds, frogs in the water hole, a distant lion, I was very conscious that my pounding heart was the loudest sound I could hear. It was growing colder, and the wind blew up under my blanket. I curled up tighter, tucking the blanket underneath my body, but I couldn't get warm and I couldn't relax. The lions continued to roar.

I contemplated sleeping in the car, but it seemed cowardly and not my style. I was not afraid of the bush, I was not afraid of wild animals. I put my head under the blanket for a while, but the lions did not go away. I finally acknowledged that by staying on the roof, I was only adding to my fear of dying alone, and realized that I would be more rational if I gave up trying to be brave and got in the car. Reluctantly I climbed down and curled up on the front seat. In the middle of the night I found my dried piece of elephant birth sac, my good luck charm; I held it in my hand until dawn.

In the cool of the early morning I worked on the car again, cried a bit, and finally lay down in the small patch of shade under the car to think. If Simon had received the message, then a search would have started by now, and I listened for planes and scanned the sky. There was nothing except the occasional jet passing 10,000 meters above. I pulled out my love letters again, reading them over and over. If I didn't survive, I would never see him again.

By late morning I began to lose hope that an aerial search had been launched, and I packed my knapsack with essentials for the walk I thought I might have to undertake—two water bottles, the rest of my food, my hat, my sunglasses, a flashlight, and my makeshift weapons. These consisted of a wheel spanner, a white plastic bag for waving in the wind, a mirror for flashing, and a tin of WD40 for spraying: an ethologist's weapons for scaring lions. Why was I suddenly afraid of lions? I had never feared them before, but Tsavo lions were different; they snarled and stared right through me, sending a chill down my spine. Everyone knew about the man-eaters of Tsavo. So I was scared of lions, something new.

There was no point in leaving the car and walking forty kilometers in the heat of the Tsavo sun. I would have to wait until the next morning. But then should I walk, or shouldn't I? For lunch I ate the rest of the croissants, a mango, and a few dates. I had already run out of water and so I drank from the water hole. I was lucky to be stranded next to it; I knew I would have water for months. But I was about to run out of food and I couldn't eat grass. I thought of the animals I had seen nearby. I had some matches. I could make a fire and roast some frogs, but then I worried that I would probably eat a noxious one, so my mind turned to insects. There were so many and I'd seen several large cicadalike creatures. Perhaps they could be roasted? I studied them closely for a while and discovered that they were cannibalizing other cicadas, which put me off.

I tried to think things through rationally. Simon Trevor had warned me that if I got stuck in Tsavo I was not to leave my car; the lions were too dangerous, the country too vast, and my chances were better if I waited for someone to find me. Before I left Lake Jipe Lodge I had asked the assistant manager to radio Simon with my plans. If she hadn't remembered or hadn't managed to get through, then it would be extremely unlikely that anyone would pass this way before I died.

The two tusks I had found the day before had been just ten meters off the road and had lain there at least three years. All through the time when the rotting carcass gave off its stench, no one had ever driven by; the lions and vultures feasted in peace. Not even the poachers found the ivory. How could I expect anyone to find me? I had a feeling in my bones that Simon had not received my message.

Living in the bush one learns to be totally self-reliant and independent, and yet there are times like this when one depends on others for one's very survival. As I sat there in the hot sun waiting for assistance, I realized that I hated to ask for help. I realized, too, that the experiences of the previous few years had left me terribly vulnerable and that my defense had been not to trust anyone. I would make it on my own.

I could walk forty kilometers without any problem. The only things that kept me from leaving the car were Simon's comments and the lions. I thought of my mother and the pain she would feel losing both her husband and her eldest daughter to Africa. That night I had a fever—heat stroke. I knew that I would leave at dawn. Using wet sand, I constructed a large arrow on the roof of my car pointing in the direction I would walk, and I left a note indicating the same on the windshield, in the remote possibility that someone passed on the road. Finally I put a note in the glove compartment for my family in case I did not make it.

Just after sunrise I walked away from my car looking ahead with my binoculars and looking back over my shoulders for the first five hundred meters. Then I was on my way, walking fast and feeling calm and confident again. I arrived at the Lake Jipe ranger post five hours and fifteen minutes later, the soles of my feet covered in enormous blisters. I had seen several herds of kongoni, a herd of impala, and numerous francolins. The lions must have stared right through me.

I later learned from Simon that the lodge manager *had* sent a radio message, but Simon is very hard of hearing and he mistook my name for another guest who *had* arrived. He then informed her that I had reached Voi safely.

CHAPTER ⇒ 27

Burnt-to-Blue Ivory

SHINGO INAVAA VYOMBO, ROHO INAVAA MAMBO.
THE NECK WEARS JEWELRY, WHILE THE HEART WEARS TROUBLES.
—A Swahili Proverb

JORGEN ARRIVED IN EARLY February, several days after my return from Tsavo, ostensibly to compile data on the status of elephants and the ivory trade for TRAFFIC. We met with Iain the following morning to discuss how we might put together a proposal that would put the African elephant on Appendix I of CITES and thus ban the ivory trade. Over the years Iain had collected an extraordinary amount of data, primarily compiled from surveys and questionnaires, on the status of elephants around the continent, and he now pulled this information from his files and handed it over to us. Armed with his papers and a laptop computer, we left for Amboseli for ten days of writing.

Cynthia was in camp, too, and several film crews were scheduled to work with us in February, filming Discovery's *Ivory Wars* and David Attenborough's *Trials of Life*. In addition, *National Geographic* magazine was putting together a major feature article on elephants, and the writer, Doug Chadwick, and photographer, Bill Thompson, both needed my assistance. There was so much to do and so little time! It was a frenetic period and Jorgen and I found only a few hours to relax and watch the elephants. One evening we took a bottle of wine with us. As we sat on the ground leaning back against the front tire of my car, a female walked up and stood with her newborn and other family members not two meters from where we sat. We froze, wondering whether we should try to slither underneath the car. We decided to stay as we were, and the elephants stood calmly smelling toward us with their trunks, apparently interested but unperturbed. The sun had slipped below the horizon, and the sky was growing soft pink as darkness gathered. The elephants had turned slate gray, but their tusks seemed to glow in the last remaining light. An evening wind blew, catching the dust disturbed by the endless movement of the tips of their trunks and sending a shiver through my body. At the elephants' feet we felt so small and insignificant, and yet so very privileged.

Our ten days were over far too quickly, and in no time we found ourselves back in Nairobi with a half-completed proposal. To be considered at the next meeting of CITES, our proposal had to be submitted by a country that was a party to the convention. Our first thought was to ask Kenya to propose the Appendix I listing, and we went to see Perez Olindo. We told him of our intentions and that the deadline for submission was May 12. Was Kenya interested?

Perez is a man who speaks in riddles and innuendoes, particularly when he is being asked a question that requires a direct answer. He is never an easy man to read, and on this occasion I found it impossible to gauge his personal feelings. Unbeknownst to us, Kenya was, at that time, preparing to sell twelve tons of ivory, most of it confiscated from poachers, and it would have been impossible for the country to propose a ban on ivory when at the same time it was intending to sell a huge consignment. During our meeting it became clear to us that Kenya was unlikely to support the proposal, and we would have to start looking elsewhere.

As it happened, we did not have to search very far. Tanzania had suffered equally serious levels of poaching, its elephant population having declined from over 316,000 in 1979 to some 100,000 by 1987. Costa Mlay, recently appointed director of wildlife, wanted Tanzania's elephants on Appendix I. We planned to meet with him when Jorgen returned in April.

In the meantime I was to carry out another survey this time in Mikumi National Park in Tanzania. There I found the elephant situation even worse than what I had found in Tsavo. Thirty-three percent of the "families" I encountered were orphan groups, and another 39 percent were obviously missing adult females. It was strange to see twenty or more juveniles in a group together, elephants of all one size. Of the 445 elephants that I aged and sexed, I saw only one male over twenty-five years old. And of the adult females, a very high proportion lacked breasts, which indicated that they were neither pregnant nor lactating, a situation I had not seen before. Among the remaining older females, some 40 percent were tuskless, the survivors of a bloody war for ivory.

In late April Jorgen returned and we traveled to Dar-es-Salaam to meet with Costa. We found him sitting at his desk facing a glass case of ivory figurines in the office of the previous director, who apparently had been implicated in the trade. Costa outlined to us the current status of Tanzania's elephants and told us what, from the Tanzanian perspective, he wanted included in the proposal.

We stayed for ten days with Liz and Neil Baker of the Tanzania Wildlife Conservation Society. Vociferous proponents of the ivory ban, Liz and Neil were wonderful hosts, providing us with music, wine, good food, and everything we might need to ensure that the proposal was completed on time. In the shade of a huge bougainvillea overlooking the sea, we bent over our computers and piles of papers, meeting every few days to discuss our progress with Costa, receive his comments, and make amendments. Over the previous few months Jorgen and his TRAFFIC team had put in an extraordinary amount of work compiling data on the ivory trade. Now we had to put the whole story together. The deadline was fast approaching and there were marathon days and nights of writing, proofreading, editing, and battling with an uncooperative printer. We completed the 150-page proposal an hour before Jorgen boarded his flight back to Washington. We had kept the true nature of our activities a well-guarded secret from all but a few trusted individuals, and we did not relax until the proposal, entitled "Transfer of the African Elephant (*Loxodonta africana*) from Appendix II to Appendix I of the Convention of International Trade in Endangered Species of Wild Fauna and Flora," submitted by the United Republic of Tanzania, reached the CITES Secretariat.

It was during this same period, April 1989, that Richard Leakey was appointed the new director of WCMD, replacing Perez Olindo. One of his first decisions in this position was that Kenya would, after all, propose the transfer of the African elephant to Appendix I. In addition, Kenya would not sell its stockpiled ivory. With Tanzania and Kenya behind the ivory ban, a wave of other countries followed suit, including the United States. In a bizarre twist, the head of the Wildlife Department of Somalia agreed to propose the Appendix I transfer without the knowledge of those government higher-ups who actively traded in ivory from elephants killed in Kenya. One by one the formerly ambivalent conservation organizations came out solidly in support of the proposed ban. In advance of the CITES meeting, Europe and the United States introduced a self-imposed embargo on the import of ivory.

In May Steve Cobb invited me to attend a meeting of the Ivory Trade Review Group in London, so that I could present the results of my surveys. Ultimately the results appeared as a chapter in a major document prepared by the ITRG for consideration by the CITES conference entitled *The Ivory Trade and the Future of the African Elephant*.

I was thrilled to have been invited to participate in the meeting, and I was blissfully happy in my personal life, too. My man was making plans to take a job in Tanzania, where I would join him, and it looked as if we might be able to move there as early as the end of the year. Deeply in love, we decided to try for a child, an effort that proved successful.

I returned to Kenya in late May only to learn that in June I would travel by way of the United States to Japan, where I was to be part of a small delegation that included Perez Olindo, David Western, and Ruth Mace, a population biologist who, with her colleagues, had modeled the impact of the ivory trade.

In Japan we met with representatives of the Japanese government and ivory traders to explain the seriousness of the poaching and ivory trade situation in Africa, and to try to convince them that a ban was the only way to go if we were to keep the elephants from commercial extinction. From Japan we flew on to Hong Kong, where we again met government officials, traders, and the press. Some of the biggest ivory traders took us to visit their shops and wined and dined us. I was beginning to feel quite sick from my pregnancy, and I remember one dinner with particular horror. As a prominent trader poured alcohol over a full bowl of live crawling shrimp and set them alight, he warned us not to try to use sentimental tactics to argue for elephant conservation as they wouldn't work. The traders simply didn't care.

We were pleased with the progress we had made in Japan and Hong Kong, but as I flew back to the United States I was aware how tired I had become. Before returning to Kenya I was able to spend some restful time with my mother at her home north of New York City, along the Hudson River. Three days before I left the United States, I watched on television as Kenya burned twelve tons of ivory. Richard Leakey had decided that the ivory Kenya had confiscated from poachers should be destroyed in a pyre for all the world to see. On July 18, as the flames consumed the remains of so many elephant lives, I wished that I had been there with my elephant-people friends, rather than watching the spectacle alone thousands of miles away. I cried for the thousands of violent deaths those elephants had suffered and for the hundreds of orphans still running bewildered through the Tsavo bush. And I cried tears of relief, for I knew that with the proposal Tanzania had submitted and with Leakey as director of WCMD, we would win a reprieve for the elephants.

Yet bitter disappointment awaited me, for on the flight to Nairobi I lost my precious baby, and two weeks after that I lost the man I had chosen to trust. Iain and Oria were at the airport to meet me and later took me to the hospital, and Saba, their eldest daughter, looked after me for several days afterward when I became ill, drawing me pictures of smiling elephants and trying to make me smile. My phone calls to my man were met at first with a deadened voice and then by secretaries instructed to say "He's not here." My letters received no response. During the long weeks that followed I imagined driving my car north for days and simply losing myself in the desert. I just couldn't believe that the sentiments we had shared could disappear overnight.

I was wrong. For reasons I will never fully understand, my dreams went up in smoke with the ivory, and I was left shattered and brittle like so many sharp-edged pieces of the burnt-to-blue ivory.

I attended the CITES meeting in October 1989, where our bitter battle for the elephants was finally won. The African elephant was placed on Appendix I and the international commercial trade in ivory was banned. The future should have looked bright, but back in Amboseli the remaining magic had gone, and my bush life only filled me with a terrible emptiness.

PART FIVE

A NEW LIFE 1990–1993

CHAPTER ❧ 28

A Voice for Elephants

THERE WAS STILL IN AFRICA A MARVELLOUS, IRRESISTIBLE FREEDOM.
ONLY IT BELONGED TO THE PAST, NOT THE FUTURE. SOON IT WILL GO.
THERE'LL NO LONGER BE HERDS SWIRLING AGAINST THE FORESTS
AND CRUSHING THEM IN THEIR PASSAGE. THE ELEPHANTS
WERE THE LAST INDIVIDUALS.
—Romain Gary, *The Roots of Heaven,* 1958

IT WAS SOON AFTER the 1989 CITES meeting that I received a
letter from Richard Leakey, asking me simply to come to see him
sometime to discuss elephants and Laikipia, a district in central
Kenya where a major elephant project was being planned. I visited
him at his new office during the first week of November, just after
sunrise. He spoke to me of what he saw as the many difficult elephant issues facing the country: not the poaching, which the ban,
he, and his security forces had miraculously already brought under
control, but the human–elephant conflict and the possible consequences for biodiversity which might be caused by the eventual
confinement of most elephants in small protected areas. They were
complicated problems requiring imaginative solutions. A strong
team of Kenyans would need to be trained to deal with these familiar elephant management questions in a new way. We also discussed the fact that a coordinator was needed for the upcoming
elephant project in Laikipia, and he asked whether I was interested.
I explained that I had spent enough time living in a tent in the

bush and I needed a change of scene. My meeting with Richard was short, and it wasn't until after I had left his office that I realized that he had been hinting at something much more substantial than the Laikipia project for me, but I would have to wait another month before I was able to speak to him again.

In late December, after a number of other possible candidates had been reviewed, I was offered and accepted the position of co-ordinator of the Elephant Program at the WCMD, which was in the process of becoming transformed into a new parastatal body to be known as the Kenya Wildlife Service (KWS). The Kenya Wildlife Service was, like the Wildlife Conservation and Management Department, to be the government agency charged with conserving and managing the country's wildlife, but as a parastatal it would have considerable independence from the Ministry of Tourism and Wildlife, and, most important, it would be able to retain the revenue it earned rather than remitting it to the government's central treasury. I would start out as a World Bank consultant working with a team of conservationists and economists to put together a comprehensive policy and five-year investment program document for the future KWS. Once the new parastatal was formed I was to stay on as Elephant Program coordinator, with my job being to set and implement elephant conservation and management policy and research priorities for Kenya and to build up a team of Kenyan graduates who could deal effectively with the many elephant conservation and management issues that faced the country. My position was to be funded by the World Bank as part of what was to be a $148 million investment program that was under negotiation, and it was to start in June 1990, giving me six months to wind up my project in Amboseli. After the euphoria and then the desolation and remorse of 1989, I was now being offered an opportunity to make a new beginning. It was also the chance that I had been looking for to step out from pure research and the confines of Amboseli.

I met with Richard again late one afternoon at the end of March 1990. He had business in Amboseli that day and had arranged to stay on for a discussion. We talked of conservation, elephants, my job, his job, my life, and his life. We discussed Amboseli, its elephant-damaged trees, the impact of tourists, and the role of the local community in conservation. We discussed the ethics of elephant culling and the possibility of the development of elephant contraceptives. We spoke, too, of the growing conflict between elephants and people, of fences and of shooting problem elephants. He warned me that there were areas of the country in which elephants and humans simply could not coexist, and that there would be places where small pockets of elephants would have to be eliminated. In other areas we would have to protect people and their livelihood by containing elephants inside protected areas with electric fences, a solution that in some circumstances could have grave consequences for a park's habitat if we were not prepared to control growth in the elephant population.

After we had talked for several hours, I asked him why he had chosen me for the job. He said he needed someone with knowledge of Kenya, someone who understood elephants, and someone who could inspire young Kenyans; he told me he hadn't found anyone else with those qualifications. I told him that I had never worked for anyone before; it was a new experience for me. He laughed and said that he had never worked for anyone, either, and it was clear from his tone that he never intended to.

Since 1988 I had rented in Nairobi small but quaint lodgings that were part of an old mews where I stayed during my short city visits. In April 1990 I locked up my various boxes of sound equipment, putting them in a corner of my tent, removed my clothes from my chest of drawers, and moved to Nairobi permanently. My tent was to remain standing as a room for guests and for my use during occasional visits. I departed from Amboseli with mixed feelings. Perhaps the greatest sense was one of being set free. But I also felt a very deep sadness at leaving the elephants and at abandoning my communication study when it was only half completed. There were still so many exciting discoveries to make, and I found it difficult to accept that I was no longer going to be the one to make them. I would miss the tranquility of camp, too, but I looked forward to the opportunity of developing a social life and close friends.

One such friendship was reestablished soon after my departure from Amboseli. I was visiting family and friends in the United States and in England when, in London, quite by chance, I met a colleague from Princeton, Andy Dobson. A man with a head of dark curls, a twinkle in his eye, and an instant smile, Andy is one of the most generous people I have known. He is a parasitologist and population biologist by training who uses his expertise to solve various conservation problems around the globe, from the epidemiology of rinderpest in the Serengeti to the demography of grizzly bears in Yellowstone. Andy's research brought him to East Africa once or twice each year, and we began to work together on a series of papers modeling the data I had collected on my various surveys to examine how the poaching might have affected the elephants' social and reproductive ability to recover.

At the end of May I returned to Kenya to take up my position at the Kenya Wildlife Service. The job quickly gave me a new perspective on the future of the country's elephants. I had to learn to use a new voice, a *political* voice, which was not always one that I believed in. I have always been one to speak my mind, and I had to learn to keep an uneasy truce with this new voice. As a behavioral biologist I had spent years advocating for the elephants from an elephant perspective. Now, as a manager, I had to stand back from them far enough to be able to see them from a human perspective, as species in a larger context.

When the decisions got tough, it would have been so much easier for me to forget that I knew anything of elephants and their lives, but I did try to retain both my objectivity and my compassion for both sides. In all the years I studied elephants in Amboseli, I never thought that I would be responsible for killing them. In 1993 alone I gave the go-ahead to shoot fifty-seven "problem" elephants. The lack of transport for our employees, their own lack of training, and the fact that elephant attacks and raids tended to take place under the cover of night meant that by the time our rangers found a group of suspect elephants, the culprit they selected was often merely a suitable scapegoat. I believe that many of the elephants shot were innocent, but politically, it had to be done. The best I could do was to ensure that I provided my team with elephant experience and instilled in them a sense of the integrity that they would need as custodians of the country's elephants.

When the decisions got especially difficult I turned to Richard. Richard is a man with vision, a man with an intuitive political sense and an almost uncanny ability to see into the future. Time and time again I walked into his office distraught at the decisions I was having to take. He laughed at me, told me to calm down, and then patiently encouraged me to stand back, take a deep breath, and look into the future for answers to today's problems. We were, he reminded me, playing a political game, and appeasing people in relatively harmless ways would prevent us from losing even more ground for the elephants. Yes, we must give better training to our men, yes, we would provide them with better equipment, but we were going to have to surrender in lots of little battles in order not to lose it all. I learned to weigh the costs in any situation and then, if necessary, to close my heart before sending a radio message to shoot another "rogue."

During my time at KWS I came reluctantly to realize that the days of freedom and wide-open spaces for the elephants were in the past. I began to see that in perhaps twenty-five years, the majority of the remaining elephants of Kenya would be behind bars. Barriers of some sort, whether in the form of human settlement or posts with wire, would surround each national park and reserve, and beyond their boundaries there would be essentially no elephants. Between now and then the kind of painful choices I had to make would be made over and over again by the managers who followed me, and in the meantime, thousands of people and thousands of elephants would suffer.

Despite the depressing conservation realities, working for

Kenya Wildlife Service was an exhilarating experience. Training a team of elephant specialists and providing opportunities for young Kenyans to study elephants gave me deep satisfaction. I eventually built a team of eleven dedicated individuals each pursuing an elephant project of his or her own, some working on master's and Ph.D.-level studies, including topics such as developing a contraceptive vaccine for elephants, examining the impact of elephants on forest habitat, improving techniques of counting elephants in forests, studying the dynamics of elephant populations and their implications for future management, looking at ways to mitigate human-elephant conflict, trying to outfox fence-breaking elephants, and finding ways in which elephants and humans could co-exist. It pleased me enormously to share the excitement I had had studying elephants and to pass on that experience to other young people.

Working with a dynamic team of Kenyans from other departments within the service, I made new friendships with some very talented and dedicated individuals, among them Carole Mwai Wanaina and Tahreni Bwana. A vivacious, intelligent, and capable young woman with easy laughter, Carole was brought to KWS as Richard's personal assistant, and we quickly established a close and lasting friendship. Tahreni, a calm, statuesque, and deeply loyal Muslim woman, born on a tiny island off the Kenyan coast and trained in the United States, ran the Computer Services for KWS. Though I worked closely with Tahreni on the elephant mortality database, we were such different personalities that it took tragedy to break through the reserve and bind us together in friendship. Yet it was Richard who still made the biggest impact on my life, teaching me to think on my feet, let go of the anger, and find the courage and the confidence I needed to do my job most effectively and to take some hard decisions in my life.

CHAPTER ⇒ 29

A Cache of Ivory

MZIKA PEMBE NDIYE MZUA PEMBE.
THE ONE WHO BURIES THE IVORY IS THE ONE TO DIG IT UP AGAIN.
—A Swahili Proverb

ON JUNE 27, 1990, four weeks after I had officially joined the Kenya Wildlife Service, I was due to meet Richard Leakey at Wilson Airport to fly to Amboseli and then on to Tsavo to join a National Geographic film crew. Instead, when I arrived, Richard hurried me into a Cessna 402 and we took off for Voi. That morning he had received a message from Steven Gichangi, the warden of Tsavo East, saying that a large cache of poached ivory they had been searching for had been located. The young poachers-turned-informers had been retrieved from Somalia by one of the KWS agents and had led our newly recruited and trained wildlife protection team to a spot west of Kasigau, on Wanainchi Ranch, where the piles of tusks lay buried.

We landed at the Kasigau bush strip and found the men waiting. The agent was a tall, thin Somali who looked as if he deserved his code name, Mamba, a deadly snake. He escorted us to two Land Rover pickups while his men stiffened to attention, automatic weapons gripped across their chests. It was a twenty-minute drive along a rough track and then a brisk fifteen-minute walk through the bush to the site.

There under a cluster of acacias were seven caches of ivory, each marked with a branch. The men had been waiting for Richard to arrive and now they started to dig. Tusk after tusk was removed, the smallest from young adult females, the largest from thirty-year-old males. The biggest cache contained thirty-four tusks. In all, we guessed the haul was close to a thousand kilograms, with an average tusk weight of seven kilograms. The tusks were weathered and stained the rust color of the Tsavo earth; they had been buried there for a long time.

I turned to study the informers, who stood silently under a nearby tree, closely guarded by our men. They were mere boys, but their young faces belied their innocence. Dressed in *mashuka,* cotton cloth tied around their slender waists, and sandals, only their Kalashnikovs were missing. Somalis. They were a tough breed who thought nothing of surviving off the bush. With an automatic weapon cradled in their arms, they could get almost anything they needed: morning tea, an evening meal, a place to sleep, or someone to sleep with. More often than not they slept in the open. The Somali poacher-bandits had terrorized the communities around Tsavo for years and held up tour vehicles at gunpoint in the late 1980s. Three foreign visitors had been shot dead, which had threatened our tourist industry and the nation's economy. "*Mlizika meno lini?*" When did you bury the tusks? I asked. "*Mwezi wa kumi.*" 1988, one answered.

October 1988. My mind drifted back to when I had driven through the adjacent Lualenyi Ranch looking for elephants in January 1989, only three months after this burial had taken place. They had been killing elephants in the same vicinity that I had been counting them. The closest I had come to the cache was on the boundary track between Tsavo and Lualenyi, not more than a few kilometers away, though it was here that I had been headed when my vehicle died.

The informers told me that they had buried the tusks because with the antipoaching forces everywhere, it had been impossible to move them out. Now, they said, "*Hakuna mtu anataka sasa. Haina bei.*"—No one wants the ivory now, it has no value. It had been a lot of work for nothing: killing all these elephants, cutting the tusks out, carrying them for kilometers, and burying them under three carefully marked trees. Then the buyers got scared and refused to pay. The KWS agent had offered the poachers money to talk, and, under the circumstances, it had seemed like a good deal.

Watching those tusks come out of the ground was like being witness to evidence of a holocaust: the slaughter of perhaps eighty elephants. One hundred fifty tusks under three small trees. What must the surviving elephants have thought when they passed by and smelled the death of their companions lingering only four inches below the surface?

I thought about the images of the elephants for a moment, and then the clanking sound of tusk thrown against tusk brought me back to the present. I watched the faces of the rangers at their work and I felt so proud of them, so proud of KWS. All of those terrible years and now, finally, some success. Some success, but what a tragic waste of life! These elephants had been slaughtered for their ivory, and their tusks buried, waiting for the right moment and the right price to sell. The right moment had never come and now the tusks would be burned. No ivory trinkets, no signature seals, just a lot of dead elephants and some ashes.

Two months later Iain and I did a reconnaissance survey of the Boni Forest and Dodori Reserve, just below the Somali border, where we hoped to undertake a full aerial survey later in the year. Iain had assisted the budding KWS Elephant Program in obtaining substantial financial support from the European Community for elephant surveys. Elephants inhabit twenty-nine of Kenya's parks and reserves, as well as most of its national forests. They are also found across large parts of the country outside of these protected areas. After the years of poaching, a top priority was to reassess the status of the country's elephants. During my tenure at KWS, most elephant habitat was surveyed and we gained, as a result, a deeper understanding not only of the elephants' conservation status, but of the management issues posed by their presence.

The Boni and Dodori area had once been home to Kenya's largest elephant population, but poaching there had been so devastating that it was uncertain whether even one elephant remained. We wanted to fly the area to see if it would be possible to carry out an aerial count. Iain, Richard, and others had told me that as recently as the 1970s elephants could still be seen sliding down the sand dunes and swimming in the sea. How I wished I had been able to witness that. Iain flew me up along the stretch of coastline from Manda Island to the Somali border. On Manda itself we counted six old carcasses where the elephants had been shot trying to cross to the island from the mainland. We continued north past the island of Kiwaiyu along a wild and romantic coastline. The sea was a beautiful blue-green, and the breakers crashed violently against the cliffs of a long, narrow deserted island several hundred meters off the mainland. Suddenly there on top of the island I saw a pile of elephant bones: the skeletons of seven elephants bleached white in the sun. Why, I wondered, had the elephants crossed the sea to this barren island? Perhaps to enjoy the view and to listen to the breakers crashing against the rocks below. Then, standing together, they had been gunned down. I almost wept at the horrible juxtaposition of beauty and cruelty.

While on that flight and on the day I stood in Tsavo watching the recovery of tusks, it may have seemed that the war with the ivory poachers was a thing of the past; it was not over. Although the number of elephants killed for their ivory declined from over three thousand each year in the late 1980s to an average of perhaps forty to fifty elephants a year from 1990 through 1993, the KWS Wildlife Protection Unit, its Intelligence Unit, and a network of informers continued to wage a war behind the scenes against the remaining poachers and traffickers. Abdul Bashir, the deputy director in charge of Security under Richard, directed all these KWS units. A powerfully built man, Abdi, as he was called, had been seconded to KWS from the GSU, and he terrified the uniformed staff, who stiffened to attention when he so much as appeared at the far end of a hall. Never having been exposed to military training and lacking all knowledge and interest in proper codes of conduct, I had a different relationship with Abdi. Despite the serious nature of his job, Abdi had a terrific sense of humor and a mischievous twinkle in his eye, and he and I shared a lot of laughs. Like Tahreni, he was a devout Muslim from the coast and he spoke impeccable Swahili. While I consider my Swahili good, Abdi loved to point out how limited my knowledge actually was, teasing me in riddles and leaving me to unravel unexplained proverbs. As head of the Wildlife Protection Unit, or antipoaching forces, a unit of several hundred men and women who were stationed around the country, and the Intelligence Unit, a team based in Nairobi charged with the collection and assimilation of intelligence, Abdi worked closely with both Richard and Tahreni. Largely on the basis of information compiled by the Intelligence Unit, 1,640 KWS employees eventually were fired, as evidence proved that many had been involved in elephant poaching and other illegal practices during the late 1980s.

Though I was coordinator of the Elephant Program, only a small portion of the intelligence gathered was shared with me. Information was passed on a need-to-know basis, and much of it was not only highly sensitive, it could be deadly. I needed to know how many elephants were killed, where, and why so that I could maintain the elephant mortality database. I needed to know the size of the gangs, the weapons they used, and the routes the dealers were following so that we could monitor the trends in the illegal trade. I needed to know whether KWS was still winning the war. I did not need to know the names of individuals involved in the trade, nor the names or faces of our informers. To know would have put both my life and theirs in danger. The incidents that were related to me had nameless characters. They were dirty stories of a dirty trade: a midnight rendezvous at the bottom of the Rift Valley, a cover blown, an informer strangled to death, a piece of wire found around his neck.

I knew that truckloads of ivory were being moved across the Kenya–Tanzania border at Loitokitok and that ivory was crossing the border into Ethiopia and Uganda; I was told of hundreds of tusks that lay hidden under a tilled field in Ukambani as our rangers, pretending to be mechanics, kept watch from a "broken-down" truck on the road; I knew that sacks full of ivory off-cuts were found abandoned on the Kitengela plains, their shape indicating that tusks were still being cut into small rectangular pieces for Japanese *hankos;* I was aware of a pile of ivory blocks abandoned on a road at the base of the Rift and of mountains of fine ivory dust at Magadi; that boxes labeled "Kisii stone," "malachite" or "wood carvings" sealed and ready for shipment were found to contain ivory covered in shoe polish. I knew of ivory found buried under a banana tree in the garden of an army colonel in Isiolo and I was aware that members of Parliament were trading in ivory.

I knew, too, that the five Koreans and one Ethiopian who were caught were just part of the trading ring that was responsible for perhaps 90 percent of the Kenyan trade. They were found with chain saws, carving machinery, and boxes sealed for shipment in a small factory north of Nairobi in Kiambu, thin disks of discarded ivory, with its unmistakable intricate zigzag pattern, littering the floor. The boxes were full of ivory, whole tusks, carved figurines, and small rectangular pieces; they were addressed to Seoul. I knew that the Koreans had tried, without success, to bribe our men for their release with millions of shillings and that they were finally able to "arrange" their freedom from higher up.

The ivory trade in Kenya was not dead but had just gone quiet, waiting for the right moment and the right price. It was waiting for a time when Kenya Wildlife Service was less vigilant, less powerful, perhaps less honest, or when the markets opened up again. The poachers and traders were just biding their time, and one day their time may come.

CHAPTER ⇒ 30

The Elephant Menace

WHAT PROGRESS REQUIRES INEXORABLY OF HUMAN BEINGS AND OF
CONTINENTS IS THAT THEY SHOULD RENOUNCE THEIR STRANGENESS,
THAT THEY SHOULD BREAK WITH MYSTERY; AND SOMEWHERE ALONG
THAT ROAD IS INSCRIBED THE END OF THE LAST ELEPHANT. THE
CULTIVATED LANDS MUST ENCROACH UPON THE FORESTS, AND
THE ROADS WILL BITE MORE AND MORE DEEPLY INTO THE QUIETUDE
OF THE GREAT HERDS. THERE WILL BE LESS AND LESS ROOM
FOR NATURAL SPLENDOR.
—Romain Gary, *The Roots of Heaven*, 1958

IN JUNE OF 1992 the radio messages began to come in from the
stations almost every day: Elephants were on the rampage. They
were eating their way through maize *shambas,* wheat fields, and
forest plantations. They were knocking over papaya, coconut, and
cashew trees. They were breaking fences, destroying dams, pulling
up water pipes, pushing over grain stores and houses. They were
preventing small children from going to school, and they were
trampling people to death. The reports came in from Kwale, Taita-
Taveta, Rombo, and Kimana, from Narok, Nyeri, Laikipia and
Rumuruti; they came in from as close to home as Limuru and as
far away as Nasolot, Maralel, and Marsabit. It seemed that wherever
there were elephants, there were problems.

In 1990 the elephants had still been too shell-shocked from the
slaughter of the 1980s to venture very far out of the protected areas.

The following year an increasing number of elephants misbehaved when the crops were standing tall in the fields and some had to be shot, but during the 1992 growing season they ventured freely into the fields and villages, and we authorized wardens to shoot problem elephants when necessary.

The newspapers headlines documented the growing conflict: "Woman trampled to death by rogue elephant"; "Jumbos on the rampage"; "Elephants a threat to human life"; "Women and baby killed by elephants." The so-called "poaching menace" of the 1980s had become the "elephant menace" of the 1990s. Peasant farmers were quoted as saying "*Ndovu ya siku hizi ni jeuri,*" the elephants of today are brutal tyrants, and they vented their anger and frustration at KWS. To their minds Kenya Wildlife Service had stopped the slaughter of the elephants by shooting poachers on sight but seemed unable or unwilling to do anything to end people's suffering. Why weren't the elephants being shot? the people demanded. They implied that KWS valued elephant life more than human life.

The situation was, of course, not that simple. As mentioned, during the previous twenty years the elephant population had declined through ivory poaching and loss of habitat from some 167,000 elephants to less than 26,000. After the wanton slaughter of elephants and the collusion and subsequent sacking of Wildlife Department staff, KWS rangers had been hesitant to shoot "problem elephants" in 1990 and 1991. With the poaching pressure off and rangers firing harmless blanks in the air, the elephants had come out of the forests and national park refuges and back along their traditional migration routes, only to find human settlements crowding their old habitats.

There were too many elephants, the people said, they had been increasing rapidly. But I knew that Kenya's elephant population could not have begun to rise until after the ivory ban had come into effect in 1990, giving them at most three years in which to have begun to recover. If there were an estimated 25,000 elephants in 1990, a 4 percent growth rate would have resulted in approximately 28,000 elephants in 1993, an increase of 3,000, but still far

fewer than were alive twenty years earlier. (In fact, the 1994 estimate was 23,000—not necessarily a decline but the result of better
surveys.) Yes, the elephants were beginning to increase, but the real
problem was that while Kenya's elephant population had been declining, its human population had doubled, increasing from some
15 million in the mid-1970s to almost 27 million by 1993.

It was not helpful, though, to explain this to a woman with a
family to feed whose only source of food had just been demolished
by elephants. It was not helpful to a woman whose husband had
just been trampled. Imagine having your entire livelihood destroyed
by a beast that comes in the dead of the night. Imagine defending
your homestead from a monster weighing close to a hundred times
your weight, one that knows exactly where you are through its
extraordinary sense of hearing and smell, while your only useful
sense is reduced to what you can see in a dim circle of light from
a flashlight with failing batteries. Imagine being too poor to buy
new batteries.

At KWS we had to work with certain givens: The human population would continue to increase, and human settlement would
continue to encroach upon elephant habitat. The underlying problems would not change the fact that elephants were considered
wildlife enemy number one and that it was Kenya Wildlife Service's
responsibility to prevent damage and injury to human life and property.

Our Elephant Program surveys had always included an investigation into human-elephant conflict around the forests and other
protected areas, and we were well aware of the problem areas and
their underlying causes. The pressure for land in Kenya was intense,
and sensible land-use policy was often abandoned in favor of poor
land-use practice. In many places there was simply no land-use
planning at all or, if there was any, it wasn't followed. Corruption
in high places exacerbated the problem. State land in some elephant
habitats was being handed out in return for political favors to individuals who then sold it for settlement. Refugees were being
resettled along a section of the Mau forest that we knew from our

surveys to have the highest elephant density. And large tracts of protected national forest were being degazetted and given over for settlement, creating "island farms" within forests and "forest peninsulas" extending into prime agricultural land. As a result maize fields pressed up against the forest boundaries, tempting the elephants to make nighttime raids. The pattern was clear along the forest boundaries of Mount Kenya, the Aberdares, and Shimba Hills. Decisions that affected wildlife were being taken by other ministries without consulting the wildlife authorities, and without any consideration for the future.

In the rangelands the conflict was equally intense. Land-buying companies purchased large ranches in Laikipia and then sold them off, sight unseen, to landless people as barren five-acre plots. Elephants became trapped in the remaining fragments of forest in Rumuruti and Ngare Ndare, or on the "ele-friendly" ranches of Sangare, Olpejeta, Lewa Downs, and Olari Nyiro—mere islands in a sea of hostility.

Elsewhere traditional pastoralists were encouraged to abandon their seminomadic lifestyles, settle in one place, and grow crops. The Rendille, Somali, Boran, and Samburu in the north of the country, and the Maasai in the south, who had once lived in harmony with elephants, now demanded their removal. Group ranch land in Narok and Kajiado was being divided into individual land holdings and often sold in small parcels to agricultural tribes, who moved onto these marginal patches and tried to grow maize there.

Elephant habitat everywhere became human habitat; traditional migration routes were blocked by settlement, while irrigation caused rivers downstream to dry up and seasonal mud wallows to disappear. As the elephants moved out along their traditional migration routes from Mount Kenya or the Aberdares to Laikipia, from Kilimanjaro to Tsavo, from Mount Marsabit to the surrounding desert, each year they encountered freshly tilled earth. The days when elephants descended from the mountains in the wet season to gather in huge numbers on the plains to breed and to socialize were over.

The elephants were now expected to stay in the land set aside for them: the national parks and forest reserves. But if it is hard to convince people to obey the law of the land, it is even harder to convince an elephant. While the people continued to insist that KWS do something about the "menace," the elephants continued to make their nocturnal raids. And when the people felt that KWS was not doing enough, politicians entered the fray, encouraging them to take the law into their own hands and kill the elephants.

As the conflict raged on, KWS documented the toll on both sides. In 1990 nine people were killed by elephants, in 1991 the figure was twenty-four, but by 1992 it had increased to forty. Though part of the increase merely reflected more efficient data collection and recording, the deepening hostility was real. The number of elephants killed in the conflict, including those shot by our rangers, also rose with each passing year, as did the number of elephants with wire snares around their legs and trunks, spears through their guts, or suppurating wounds from poisoned arrows. One elephant, snared and unable to walk, fell into the Mara River and lay there, underwater, for three days, holding the tip of his trunk above water in order to breathe. Another, a newborn baby, left a foot in a snare in the forest of the Aberdares. The wounded who were found by the KWS vets were treated or euthanized; those who were not died a long, slow, and painful death from starvation, septicemia, or poisoning.

The Elephant Program sent teams to the field headed by Winnie Kiiru, Moses Litoroh, James Sakwa, Steve Njumbi, and others to investigate the causes of the problems and to recommend solutions. Entire projects were focused on solving the crises in Laikipia, Taita-Taveta, and the Shimba Hills. With KWS's Community Wildlife Service we searched for ways in which people and elephants could live in harmony through revenue-sharing or revenue-generating projects. Where elephants and people could not coexist we erected fences, we shot elephants, and we drove elephants out of troubled areas with helicopters and bands of armed rangers.

Under the KWS Protected Areas and Wildlife Services Project

(funded by the World Bank and bilateral donors), 1,500 kilometers of wildlife barriers had been planned. At the cost of tens of thousands of dollars, we built and tested a number of different experimental elephant barriers, but the elephants always seemed to be one step ahead. They caved in ditches, pushed over stone walls with their heads, and used their hind feet to knock over the posts supporting the electric fences. They used their tusks to snap the live wires of electric fences, or they shorted them by simply pushing the live wire onto the ground wire. At vast expense we tried sawing the tusks off these wire-snapping elephants, only to find them still on the wrong side of the fences and still with maize kernels in their dung. Some elephants were even known to pick up logs and throw them at the wires to short them out. In the Aberdares where four notorious males forced their way out through the main park gate time and time again, extra wires and extra strength was built into the fence. The elephants were so annoyed that they used their tusks literally to plow up the road in front of the gate. Eventually they found their way out somehow and returned to raiding around the town of Nyeri. We learned through trial and error that if an elephant wants to get through a barrier badly enough, he or she will find a way. In the end I was forced to conclude that we might have to shoot some offending fence-breakers as a lesson to their companions. But still, in order to teach elephants not go through the fence, the culprit had to be caught in the act. How could we hope to monitor miles and miles of fence line and catch elephants in the act of fence-breaking in the dead of night? The number of people, vehicles, and night vision equipment such a program would require to be carried out successfully around the country was beyond our capability at the time.

But we had to make a start, and our solution was Problem Elephant Control, or PEC. "Problem elephant" was a phrase that was used to designate repeat offenders of certain elephantine crimes: fence-breaking, crop raiding, the killing of livestock and humans, to name a few. "Control" referred to a variety of measures undertaken to reduce the number of elephant incidents; all too often it meant killing elephants. The numbers of elephants shot on control by our own men increased each year: ten in 1990, fifteen in 1991, forty-two in 1992, and fifty-seven in 1993.

I knew that in many cases the wrong elephants were shot and that far too many took too long to die or were left merely wounded. Our rangers lacked transport and training and were, in many cases, working with World War I and II 303 rifles that frequently jammed. I felt strongly that if we were going to shoot elephants, we must ensure that the men were properly equipped and trained. I wanted to ensure that they knew how to identify individual elephants, track them, and shoot them effectively. In November 1992 Norah and I worked with twelve men, the beginnings of the Problem Elephant Control team, teaching them to age, sex, and identify elephants and to instill in them a respect and an understanding for the animal that they would have to kill.

These men went on to receive further on-the-job training, learning ultimately how to kill elephants effectively and humanely. I set up a reporting system that had to be submitted for each elephant shot. With time our records improved, but still I was unhappy with the number of cases in which the wrong elephant was killed and how long many of them were taking to die. I also remained unconvinced that the shooting was having any measurable effect. How many elephants would have to be shot throughout the country every year to ensure that they were kept "under control"? If fifty-seven deaths were inadequate to alleviate the situation in 1993, would the effective offtake be sustainable in the long term? And how would the continual harassment of elephants by our rangers and others affect the elephants' behavior? I feared that if the crisis was not handled properly, we would end up with increasingly aggressive elephants and more people being killed than ever before.

The public tended to forget that KWS was a young organization. Financial support for the Elephant Program provided from the European Community had not yet started to flow, and we had only one vehicle to deal with the growing crisis. The Community Wildlife Service and the district wardens, too, had little or no transport. We were stretched to our very limits.

I am confident that with time and with the cooperation of all of those with experience with elephants, we will find solutions to many of the human-elephant troubles we are now facing. The problems are not insurmountable, but they are certainly more formidable than many people believe. Some of the solutions to today's problems will only create further dilemmas tomorrow. For example, fencing in growing populations of elephants will in many cases lead to the loss of biodiversity in small protected areas. Then what will the management answer be? In southern Africa the solution is culling, the killing of a certain percentage of the elephant population each year. I personally find the culling of elephants ethically unacceptable, and under my program we carried out research to develop a contraceptive vaccine for elephants instead. Will it ever work, and if so, will others consider it necessary or practical? In some areas there will simply be no money for fences, or they will be deemed inappropriate, and final solutions will have to be called upon: the extermination of whole populations of elephants.

CHAPTER ⇒ 31

A View Over the Rift

THERE WAS NO FAT ON IT AND NO LUXURIANCE ANYWHERE;
IT WAS AFRICA DISTILLED UP THROUGH SIX THOUSAND FEET,
LIKE THE STRONG AND REFINED ESSENCE OF A CONTINENT.
—Isak Dinesen, *Out of Africa,* 1938

SOON AFTER I JOINED KWS in June 1990 my camerawoman friend
Barbara also moved to Nairobi, and we chose to share a small house
on the outskirts in the residential area of Karen. We lived together
on the edge of the Ololua Forest in an old corrugated iron shack
built on stilts with forest-green paint peeling off the metal roof and
sides. It was not an attractive structure, but since it had been Karen
Blixen's one-time coffee store it qualified as quaint, a relic of *Out
of Africa.* Renovated many years earlier, first as a dance hall for the
local community and later as a cottage, it had old wooden floor-
boards and overlooked the Mbagathi River and the forest. It was
the last house on Karen Road, just past the Karen Blixen Museum
and before the small bridge that led into the forest, where we
walked our dogs in the evenings. The forest was still home to many
of the animals that Karen Blixen had written about: Sykes monkeys
alarmed at us from the safety of their trees, duikers and bushbucks
hurried into the thick undergrowth, a pair of rare crowned eagles
nested in the tall forest tree next to the waterfall, and we sensed

the silent presence of leopards in the dappled light under the olive trees. Barbara had once walked right into a lioness who had been eating her way through our neighbors' horses and cows.

It was an ideal place to live, ten minutes from Kenya Wildlife Service, five minutes from the Karen shops, and close to the homes of many of my friends, but after a while I began to grow restless. After years of waking to the sounds of the elephants and a view of Kilimanjaro, I needed more space, a place away from the crowds, where the air was sweet and there was a view across the open plains. I needed a place that I could really call home.

For so long I had lived in Kenya from one year to the next, considering it my home, but always conscious that someday I might have to leave. The tentativeness had been unsettling, and now I felt the need for more permanence and stability in my life. For years I had put off making concrete plans for my future because, it had seemed to me, "settling down" was something one did with a partner. Equally, I thought that by making my own plans I might limit my chances of finding a partner. But the relationship in 1989 had left me so shattered that in my thirty-fifth year it had become painfully clear to me that I could well wait forever. I decided then that I would buy land and build a house on my own.

I had seen a property that I liked, overlooking Nairobi National Park, and I needed advice. Was it a good investment? Would it be safe for a woman alone? I had turned to Richard Leakey. His answer was no: It was too close to the growing town of Ongata Rongai and, in his opinion, not a safe place for me to live on my own. But he invited me to come and look at a piece of land near his home on the edge of the Great Rift Valley. It was there that my brother, Bob, who was visiting, and I first went one day in January 1991. Richard and his paleontologist wife, Meave, walked us first to a beautiful site to the north of theirs and then to a dramatic piece of land immediately to the south.

We stood on the rocky outcrops along the fence line at the top of the property, at the top of the escarpment. The view was breathtaking. From where we stood the Rift Valley continued north to

Syria and south to Malawi. The view of the eastern escarpment to the north was blocked by the swell of the Ngong Hills, but to the south the escarpment continued as far as the eye could see, past Lake Magadi, Lake Natron, *Oldoinyo Lenkai,* and the Ngorongoro Highlands. To the west, across the valley floor, lay the western wall of the Rift, the Mau Escarpment, and, farther south, the Loita Hills and the Nguruman Range. In between, along the valley floor lay strewn the many extinct volcanoes from north to south: *Longonot, Suswa, Olegesaile, Shombole, Gelai,* and then *Lenkai,* the Mountain of God, the active volcano. Turning around, the view over the crest of the escarpment took in Mount Kenya, the Aberdares, and Kilimanjaro.

Richard and Meave left my brother and me to explore. We scrambled down the steep rocky slope to the cliff, *Esoit Naibor,* as the local Maasai called it, the white rock, where we could see an acacia forest and a dry riverbed meandering through the narrow valley over 300 meters below. Along the edge of the river we could make out patches of emerald green where the water seeped out from the hillside in a series of springs, providing the Maasai with the water they needed to survive the long dry season. A white ribbon of goats descended a well-worn trail on the opposite hill at a run, and the sound of cattle bells, of Maasai women washing their clothes, and of children singing floated up to where we stood. It wasn't hard to imagine that it had remained like this, almost unchanged, for centuries. And though the human population in this area would continue to grow, the view could never be spoiled. This was the Africa I loved, the great open spaces in soft, earthen colors. The very edge of the Rift Valley escarpment: to live here would be right for me, I thought.

I had fallen in love with the site, but contending with the problems of building a road, getting water from the spring at the bottom of the valley, and constructing a house on a twenty-five degree slope full of enormous rocks would, I knew, be impossible for me. Richard just looked at me and smiled. "I'll help you," he said. "That's a promise." I looked back at him. This man who was already so heavily committed as chairman of the National Museums of Kenya and director and chairman of the board of Kenya Wildlife Service was offering to help me develop a piece of Africa and build a house. I had heard many promises in my life; perhaps this was just another. Over the following months and years I learned that Richard was a different sort of man, a man of his word, a man of integrity and incredible generosity. I learned that Meave, too, was a remarkable woman. A private person, Meave is in many ways a true Leakey. A woman of tremendous energy, loyalty, and inner strength, she cares deeply about making the world a better place. She is devoted to her paleontological work at Lake Turkana, to Richard, and to their two daughters, Louise and Samira. As a family they are a lively and stimulating group of individuals to be with, and over the coming years I enjoyed getting to know them all.

Several days after our visit to the Leakeys', JB Burrell, my brother's tall, lithe girlfriend; Andy Dobson, who was on one of his East African field trips; my brother; and I left for the Serengeti. A month before, my brother and I had spoken late into the night about our father, a discussion that had forced me to face the events surrounding his death. Our safari to the Serengeti was to be a special visit to the site where we had left our father's ashes. It was twelve years since we had taken them to the Moru *kopjes,* and I hadn't been back to visit him. The emotions had been too close to the surface, the pain too deep. With the help of my brother I had begun to strip away the layers of denial and come to terms with his being gone.

The Moru *kopjes* are a cluster of perhaps twenty to thirty large, rocky outcrops, islands jutting up out of the flat expanse of the Serengeti plains. We had no map to indicate on which *kopje* it was that we had left our father's ashes all those years ago; we simply had to locate it by memory. In two cars we stopped every kilometer or so to scan the landscape with our binoculars, searching for something familiar. It is strange how the mind works, and I wonder what it was about one particular *kopje* that made my brother and I recognize it at exactly the same moment. We both jumped out of our respective cars and shouted, "It's that one!" With our binoculars we looked up at the top of the *kopje,* where we remembered leaving the plaque, and indeed it was still there.

Dusk had fallen and we made camp nearby. We would climb the *kopje* the following morning. Dawn broke with the clear air and brilliant sunshine of January and my father's favorite line from *Out of Africa* came to mind: "In the highlands you woke up in the morning and thought: here I am, where I ought to be." How he would have enjoyed being with us that day! We walked the kilometer to the base of Daddy's *kopje* and scrambled up the sides. It was not an easy climb, and we had to negotiate some difficult traverses and assist one another up the steep sides. Finally we reached the top and it was just a matter of making our way, scratched and panting, through the tangle of vegetation. The birds were singing, and the wind was blowing strongly from the east. Swifts and eagles soared overhead and butterflies flitted around the base of Daddy's rock. The plaque was still in its place and suited him as well as it had twelve years before. It would need some new cement soon and perhaps it could do with a polish, and I imagined that the next visit would be with my sister and my mother. But it had survived the sun, the wind, and the rain well.

ROBERT KEYES POOLE
1932–1978

He will hear the birds sing
And the hyenas call
He will feel the thundering hoofbeats of the wildebeest
The sun's warmth will pour down on him
And the wind and the rain
Will make him a part of the earth he loves.

As my brother so rightly pointed out, his ashes had mixed with the earth long ago, and he was already part of the plants growing up the base of the rock. If asked, my brother and I would say that we, like our father, are not believers in the presence of souls or an afterlife, but we had the strong impression that morning that we sensed his presence there. It felt good, and I grew calmer and more at ease with the world than I had been in a very long time.

Soon after the Serengeti safari I took a trip with Richard to Amboseli. He had a number of park management issues to discuss with the warden, and we used the opportunity to go over some elephant-related matters, as well. We were to stay in camp and leave early the next morning for Nairobi. My visits to Amboseli were far too few and short during the period of my KWS job, and I cherished the opportunity to sleep again in my own bed, in my own tent under the stars. In many ways I hated my rushed Nairobi existence with the incessant ringing of the telephone and people knocking at the door. I often longed for the peace and tranquility of an Amboseli day, the long hours of solitude with only Norah and the elephants for company. But it was always strange to return to Amboseli, and I felt the familiar ambivalence and sadness pulling at my chest. In part I felt something close to revulsion: the billowing dust, the unbelievably bumpy roads, the loneliness, and almost heart-of-darkness atmosphere that had, for me, characterized my last few years there.

We were airborne by seven-thirty and, as is always the case with Richard, we were well ahead of schedule. As we left the familiar Amboseli landscape behind I found myself reliving all of my mixed feelings. I thought that Richard must have sensed my unease and silent reflection, for he said, "We aren't due at Wilson Airport until nine. Let me fly you over Natron. It will be beautiful in the early-morning light." I had never seen Lake Natron before and I had no idea what to expect. I did not realize then what, in retrospect, I understand: Richard's suggestion was not an impromptu one; he had planned to take me there for a reason.

It is not easy to have a conversation over the loud drone of a aircraft engine and I was left to my own thoughts. *Oldoinyo Orok*, the Black Mountain, which lies due west of Amboseli, was soon before us and prompted me to reflect on the elephants, their future, and what I, as coordinator of the Elephant Program, might be able to do to make a difference. It was here, in this mountain forest, that rhinos and elephants once roamed. The Somali traders of Namanga made their living smuggling ivory and rhino horn from Tanzania's *Longido* plains across the border into Kenya, and by all accounts the elephants and rhinos of the Black Mountain went extinct in the mid-1970s, about the same time that I first went to live in Amboseli. I had been told that up until the early 1970s, the Amboseli elephants had once moved freely from the Selengei River in the north, across the *Longido* plains to the south, from the Chyulu Hills in the east, to *Oldoinyo Orok* in the west. During the rains, the elephants gathered in large aggregations of several hundred individuals and, guided by one of the great matriarchs, they began their trek northward to the Selengei. The trail they followed had been used for decades and was so wide and so deep that cartographers had mistaken it for a road. During the first wave of ivory poaching in the early 1970s, the elephants abandoned their migrations to the Selengei, but the trail could still be seen from the air as a dotted line of acacias that had flourished when water captured in the deep path allowed seeds embedded in the layers of dung to germinate.

The poaching for ivory reduced the Amboseli population from just over a thousand elephants in the late 1960s to less than five hundred by 1978. The gleaming white skulls of the dead had been the only reminders of the vast area they had once roamed, and only those who had sought refuge inside the small national park had survived. It was during this period, late 1972, that Cynthia first began to photograph and catalog the individuals and families of Amboseli. By the time I had joined the project three years later, hundreds of elephants had been killed. It was many years before we realized the extent of the slaughter, piecing the clues together from aerial surveys, carcass ratios, the many photographs of missing elephants, and the memory of eyes rolled back in terror.

The Amboseli elephants had escaped the second wave of poaching that devastated so many of Kenya's populations in the late 1980s by staying within a few kilometers of the park where the high density of tourists and vigilant researchers had provided protection. But now their sedentary lifestyle had become a threat in itself. As the population continued to increase, the acacias could no longer regenerate and some claimed that biodiversity was declining. I knew that if the elephants did not return to some of their old migration routes, Amboseli itself was doomed. It could not survive the continual presence and pressure from so many of the earth's largest land mammal. There were louder and more frequent calls to "do something" about the elephants. "Doing something" was usually a euphemism for culling, and culling just an easier way of saying killing, but I recognized that if I could not come up with another option, that might indeed be what would happen.

Now as we flew westward toward *Oldoinyo Orok* I thought to myself, if *only* I could persuade the elephants that it was now safe outside the park. If *only* they would reestablish their migrations north to the Selengei, to *Oldoinyo Orok* and the *Longido* plains. Finding a way to encourage them to spend more time out of the park was a dream I strove to fulfill.

There were some signs that the elephants were moving out: Monthly aerial counts by David Western were showing lower numbers of elephants in the park than in previous years; members of the Elephant Project were having increasing difficulty locating certain groups; the Maasai Game Scouts had begun to report sightings of elephants north of *Eremito,* toward the Chyulu Hills, and across the Tanzanian border. There were also occasional reports of elephants heading toward the Black Mountain, but these were dismissed by most as unreliable. In my heart I wanted to believe that the elephants were once again walking the seventy-five kilometers to the Black Mountain. Aside from practical conservation reasons, I wanted to believe that part of the Africa I loved, the great open spaces, the freedom, still existed there. The elephants were my litmus test. If there was space for them, surely there was still the freedom for us.

The border town of Namanga was on our left, *Oldoinyo Orok* dead ahead. The Cessna 206 climbed steeply toward the mist and clouds that surrounded the rocky summit. I fixed my eyes on the ground below searching for any signs of elephants—a broken branch, a gleaming yellow twig stripped of its bark, a large gray shape, a flash of ivory. Suddenly through the mist an unexpected world unfolded before me. All these years the eastern wall of the mountain had protected a secret valley from my view. Here deep gorges, waterfalls, open glades, and a thick green forest were hidden in a steep-sided valley in the very center of the mountain. I turned to Richard and smiled, we were thinking the same thought: Here was an undiscovered gem that we must find a way to protect. We continued to climb toward the meadows and granite cliffs of the western summit. Just below the huge rocky outcrop a stream meandered through the meadows and down over the rocks, descending in little waterfalls from pool to pool. Then abruptly as it had appeared the vision was gone as sheer cliffs plunged 1,250 meters into the dry bush country below.

There were no roads here, and only the narrow cattle tracks crisscrossed the dry plains. A few Maasai settlements dotted the otherwise flat, desolate landscape. I looked ahead as Richard pointed out the two mountains that bordered the southern end of Lake Natron: *Gelai* and *Lenkai.* As we rounded the slope of *Oldoinyo Gelai,* one of the most breathtakingly beautiful sights in the world met my eyes, a kaleidoscope of colors, patterns, and movement, and changing horizons: the intense pinks of the lake, the soft blues of the mountains, and the millions of wing beats of the flamingos. As I watched the shifting light and colors, I found myself speechless, witnessing the exquisite beauty of Africa in juxtaposition with so many bittersweet memories. It had been so many years since I had allowed myself really to feel the beauty in anything. Over the drone of the plane I heard Richard ask, "What are you thinking?" "It is hard to describe," was the best I could answer, for I knew that to say more would mean tears and painful explanations. But even those words proved too much, and the tears rolled down my cheeks. I stared hard out the window, hoping he wouldn't notice, but of course he did. He asked why I was so sad, and I had to explain that actually I was incredibly happy. Of course, that in itself required further clarification, and so I tried to describe how I hadn't been able to appreciate the beauty of Africa in so long and that I had been trying to discover my passion for life again and could not seem to find it. And as the Ngong Hills and my rocky slope drew nearer on our flight path, I told him of the incident on the Ngongs and the death of my father, and I alluded to other more recent losses.

The Serengeti safari and the flight over Natron restored my spirit. And sitting on my cliff on the edge of the Rift feeling vertigo, I knew I would buy this beautiful piece of land that seemed to reflect a bit of my character. To live there would be "to have my head turned," as Richard would later say. To live there, with the Leakeys as neighbors, would be a life of challenge, contrasts, and changing colors.

CHAPTER ⇒ 32

The Birth of Selengei

UCHESHI WA MTOTO NI ANGA LA NYUMBA.
THE LAUGHTER OF A CHILD LIGHTS UP THE HOUSE.
—A Swahili Proverb

THE SUN WAS JUST beginning to set over the western escarpment when I arrived with my last load of furniture for the day. I made my way carefully down the steep rocky slope and up the ladder onto the wooden deck. The distant range of mountains were a deep purple silhouetted against the fiery orange of the sky, while the hills across my little valley had turned a soft mauve, dotted with the darker patches of acacias. On the horizon, by the edge of Lake Magadi, *Shombole* and *Gelai* stood out sharply against the fading light. The rains must be near. I stepped inside, quickly filling the lanterns with kerosene, lighting them one by one. The room, now decorated with wood, brass, and Oriental carpets, filled with a warm glow. I looked out on the view again. From inside the house the land dropped away so sharply that the distant mountains were my closest point of reference. Most houses give one a sense of being well and truly grounded to the earth. This one was different, and made me feel as if I were floating over the Great Rift Valley. I was, in a sense, flying, soaring with the eagles.

Esoit Naibor, the White Rock, September 20, 1992. Though each sunset was beautiful, there was never another quite like that on the first night in my own home, the one I had designed and Richard had helped me to build. He had warned me that the experience would be more binding than a marriage. Buying a piece of land and building a house was "really getting your anchor stuck in the corals," as he had expressed it. The construction had been quite a project, but with his laughter, advice, assistance, and constant reminders of what needed to be measured and ordered, it had been done. The first task had been building a water tank next to the spring near the bottom of the valley, which involved transporting sand, cement, and quarry stones down the steep slope on donkey back. Solar power would be used to pump the water up 215 vertical meters to another tank on the top of the escarpment, from which it was gravity-fed down to the building site. Carving the driveway out of the hillside had been the next project. It had been no simple task, since my boundary sat at the end of the last bit of flat ground. As the driveway was being built a Stone Age pot was unearthed, which Richard's mother, Mary Leakey, estimated to be some four thousand years old. Finally, the house had been built, nestled in against the side of the hill, down over the edge of the escarpment.

I sat on my wooden deck, my legs dangling over the edge, watching golden-winged sunbirds flit among the orange Leonotis. I was making a list of birds in my "garden," which dropped over 300 meters from where I sat to the valley floor below. Here, on the edge of the Great Rift Valley, I was kilometers away from the noise and smells of Nairobi and, at over 3,000 meters above sea level, almost 450 meters higher than the capital. At night and in the early mornings I felt the fresh cool air of the highlands, but the afternoons brought the dry, burning heat from the valley floor.

The corrugated iron roof peaked just above the crest of the hill, catching the gale that raged above it. At night the wind typically howled, but the mornings were strangely quiet, as if the whole valley was covered in a warm blanket. The sun rose behind the house, over the top of the escarpment across the Athi-Kapiti plains, and it was several hours before the direct rays were upon us. On clear mornings the far hills changed slowly from mauve to pink, and from my upper tank, I could see both Mt. Kenya and Kilimanjaro. On gray mornings the mist poured down over the top of the escarpment past my bedroom window and over the cliff below. I had no electricity, no telephone, just the vast expanse of Africa before me. It wasn't so very different from Amboseli, except that I was no longer alone. Within me a tiny life was beginning.

As a young woman I had always dreamed that I would meet a man with similar interests and that we would lead an exotic life raising a family in the bush. Watching elephants provided long hours of solitude and, for one with a vivid imagination, offered a fertile landscape for dreaming. Left alone to my many fantasies, I was an Amboseli dreamer. However unrealistic, dreams have a way of capturing one's imagination, but they lead to a deep sense of loss when they are shattered. At thirty-six I had already mourned my lost opportunity for love within a nuclear family until I no longer wished things could have been different. I had looked love straight in the eye and I accepted that, for me, fulfillment would come in the form of a child, not a marriage. I knew from past experience that there was no reason why the route I was choosing, though difficult, couldn't be just as rewarding as my original hopes. I was very much at peace with my decision. I owned land, and I had a place to bring up my child with a breathtaking view over the earth.

As I sat watching the sunbirds dip their beaks into each of the fiery red flowers, it seemed odd that making the choice to have a child on my own had caused me such anguish. For months I had gone over the various options for getting pregnant with my friends until I had received the full range of advice for a single mother: the anonymity of a sperm bank; the apparent simplicity of just getting pregnant somehow, somewhere, with some anonymous male; the possible future complications of having a child with a trusted friend; the emotional fulfillment of carrying and raising the child of a man you love; the anguish associated with bearing the child of a man you love but cannot acknowledge as the father. I listened to the counsel and considered the various possibilities. My scientific mind objectively analyzed the pros and cons of every option while my heart ached. It was, in a strange way, the most difficult few months of my life, but I finally made my decision, and I know now it was the right one.

I knew that I could not cope totally alone, especially during the first year. Pauline Wathoni, a wonderful, warm, and kind-hearted woman, looked after my household, but I would have to continue working, and I needed someone who could accompany me to the Nairobi office to look after my baby. I could think of no better person than Norah. She had just had her second son, naming him Iain after our mutually favorite elephant, who was himself, of course, named after a person. Norah was one of the first to learn that I was pregnant. Would she consider leaving Amboseli to help me? Later on the day I posed the question she told me that she was ready to leave Amboseli, at least for a while, and that she and Iain would move to *Esoit Naibor* with me before the baby was born.

May 1, my thirty-seventh birthday, came and went accompanied by dancing until the small hours of the morning at Barbara's house. Then, during the early hours of the seventh of May, the contractions began. I rode their waves through the day, thinking of the deep, slow breathing of a sleeping elephant, the rhythmic pace of their footsteps crossing the salt pans, and remembering Richard's and Meave's comment that women have been giving birth for millions of years. I went through labor and the birth of my baby as an elephant would, surrounded and assisted by my female companions: Cynthia, the wise matriarch, was at my side once again, as were Norah, the allomother, and Barbara, the female bonded supporter. And when at last my child came into the world, she arrived like a baby elephant, amid great commotion and ceremony. Stretching her arms wide, she seemed to embrace it all.

I named her Amber Selengei. Amber was for the deep russet richness of the elephants' eyes, Amber for the essence of a continent distilled through millions of years, Amber for timeless beauty. Selengei was a young maiden of romantic Maasai legend, a girl turned to stone for love, who stands on the banks of the Selengei River, once the destination of the Amboseli elephants during the rains, before the poaching began.

Three days later I returned home. I slowly walked around my little house on the edge of the Great Rift Valley, and it seemed strangely different to me now, its colors brighter, its smells sweeter, its rooms welcoming, beckoning me closer. To reassure myself that everything was as I had left it, I slowly opened each of the kitchen cupboards and studied their contents. But here, too, things were not as I remembered them, and I said aloud, "I feel a stranger in my own home." Pauline smiled and said, "You are. You left here as one person, you have returned as another. Now you are a mother." So that was the explanation; I needed no other.

In the days, weeks, months that followed, the washing was end-less, and I frequently apologized to Pauline. Her answer was always the same, "*Hii ndiyo ile kazi tulikua na ngojea, kazi ya Selengei.*" This is the work we were waiting for, the work brought by Selengei.

When Selengei was two weeks old we held a Maasai ceremony, *Emasho Enkerai pee Edungokini Enkarna,* to cut a name for the baby. According to Maasai tradition, babies are referred to by pet names until they are several weeks old, at which point they are given their real names. The evening before Ole Nemushai, Richard's body-guard, brought a goat and a sheep, which were slaughtered at the top of the hill and hung overnight in the store. The following morning he made *motori e tomononi,* a special soup for the new mother, from the goat's neck bones and, to give me strength, he added *Olchani* from the bark of the *Acacia drepanolobium* tree. To feed the many guests Pauline prepared several coastal dishes, *Pilau, wali wa nazi, mchicha,* rice with lamb and cardomom, coconut rice, and greens with coconut, as well as forty-three chapatis.

Late in the afternoon Maasai women neighbors arrived bedecked in colorful *mashuka* and beads. To my surprise and delight, Norah and Pauline had borrowed my Maasai jewelry and *mashuka* and joined in the procession of women. They arrived blessing the baby, gently spitting toward her to prevent the evil eye, and softly exclaiming *"Oh pasinai!,"* Oh my trouble!, a term of endearment that I had heard for many years but was only beginning to appreciate. They sat in a row along one end of the deck, Selengei in the arms of the matriarch, and the ceremony began. They repeated several names that were not her real names, to which I gave no response, until finally they spoke the name Selengei, to which I answered, *"Eyioo,"* Yes, *"esere,"* it is good. The leader of the women handed Selengei to me and, as I sat on a low, three-legged stool, they sang and danced around me. A dark cloud followed by a sheet of rain advanced up the valley toward us. The long rains had all but failed and several women now commented that this unusual storm was surely a sign of good fortune for Selengei.

I am single and now I am a mother, a combination that makes a difference to some, not to others. In my own Western society the issue for many is that I am a single mother. For the Maasai it is whether I am blessed with children or not. The Maasai ask first, *"Keningai?,"* To whom do you belong? I answer, *"Ene Poole,"* The daughter of Poole, or *"Maata Olmoruo,"* I do not have a husband. The latter answer is cause for lament. The second question is always, *"Kaata inkera?,"* How many children do you have? My answer has always been *"Maata inkera,"* I do not have children, to which they exclaimed, *"Ushoo!"* for this is the worst condition for a woman to find herself in. Now I reply *"Ata nabo,"* I have one daughter, and they say, *"Esidai,"* It is good. It is understood that, for a woman, life is not complete without this passage. It is part of a coming of age. I am no longer known as Joyce, I am *Mama Selengei,* the mother of Selengei. *Esidai,* it is good, so very good.

CHAPTER ⟫ 33

A Family Reunion

THE WRENCH WOULD HAVE BEEN LESS HARD IF I COULD HAVE GONE
UP TO THEM AND SAID GOODBYE. BUT IT WAS ONLY IN DREAMS THAT I
COULD DO THAT. THEN I WENT AMONG THEM AND EXPLAINED ABOUT
THE DETESTABLE CLICKING MACHINE, AND THEY TALKED TO ME AND LED
ME THROUGH WONDERFUL FORESTS. THEY CARRIED ME ON THEIR BACKS
AND ALLOWED ME TO WATCH THEIR MYSTERIOUS DANCES AND
TO PLAY WITH THE LITTLE CALVES.
—Vivienne de Watteville, *Speak to the Earth,* 1935

AFTER MY FATHER'S DEATH my mother had told me that she would
not return to Africa. With Selengei's birth she had prevaricated for
months until finally, when Selengei was seven months old, she
came to visit. We planned a safari to Amboseli that was to be a
family reunion of sorts. My brother, Bob, was making an elephant
film with me in Amboseli for National Geographic, and his girlfriend, JB, was recording the sound. They, too, would be staying
in camp. Cynthia, who had, over all the Amboseli years, played the
role of mother and sister for me, would be there, too.

The day after we arrived we went out to watch elephants. It was
early evening and my brother waited for the best light. He wanted
footage of Selengei and me with the elephants. I knew that if we
went to *Longinye* we could catch the elephants leaving the swamps
at the end of the day against the backdrop of Kilimanjaro. We
found Big George in musth at Kudu Corner, with Chloe's and

Orlanda's families nearby and several other groups moving slowly in our direction. Bobby was thrilled: He had already filmed Big George here in the same light, and footage of Selengei and me would cut nicely with his earlier work. Over the handset Bobby and I discussed the shot he wanted, and then I heard his voice "Rolling!"—my cue to enter frame. With Selengei balanced in my left arm, I drove over the tussocks toward Big George, approaching him as I always did, in a wide arc, so that he could see me clearly. I continued toward him until the change in his posture told me I had reached his limit of tolerance: three meters. As he spun toward us I switched off the engine in time to feel the deep vibrations of his musth rumble and to hear the splatter of his urine on the earth. This classic musth male threat had by now become a game for us. Big George towered over the car, his tusks so close that I could have reached out and touched them, the sweet smell of his secretions hanging heavily in the air. Selengei stood on my lap staring up at him wide-eyed, her mouth open, her chubby little hands grasping the window frame. Big George stood silently above us, only the pitter-patter of his urine breaking the stillness. But the tip of his trunk was slowly turning left and right, testing the air. I knew that we were already forgotten, his great mind returning to the scent of Chloe. He strode forward and, with his trunk outstretched, he grasped Chloe by the tail, stopping her in her tracks. He placed the tip of his trunk inside her vulva, slowly opening and closing his eyes as he concentrated on the smells he was sensing. She held still for him, a sign that she was not in estrus.

Another group was moving rapidly toward us across the *Longinye* plains. It was the Vs, Amboseli's largest family. As they approached I could see Vee in the lead, with the two unmistakable large **V** notches in her right ear. It still seemed strange not to see Victoria and Virginia, both with their asymmetrically crossed tusks, among them. Virginia had always been a favorite elephant of mine, and often when I was alone I stopped to sing to her. It was a ritual we had; I sang "Amazing Grace," and she and her family stopped to listen. Virginia would stand silently, slowly opening and closing her amber eyes and moving the tip of her trunk. I sang for five or ten minutes, or for as long as they would listen. Virginia had been the oldest female in the population when she disappeared. We never found her skeleton, and with her death, an important piece of information was lost—verification of her age, as estimated by the wear on her teeth. Her presumed younger sister, Victoria, died soon after, also in some unknown spot, perhaps submerged deep in the swamp. Now Victoria's daughter, Vee, was matriarch.

The Vs seemed to be heading toward Big George, and so I moved the car slowly forward. I wanted to be completely surrounded by elephants, to feel and smell their presence again. I held Selengei close and steered the car slowly toward them. I wanted the group to split, some passing in front of the car, some behind, and others to stop and look in the window at us. When I was directly in their path, I turned off the engine. The elephants were less than two meters away from us when Vee stopped dead in her tracks and, with her mouth wide, gave a deep and loud, throaty rumble. The rest of the family rushed to her side, gathering next to our window, and, with their trunks outstretched, deafened us with a cacophony of rumbles, trumpets, and screams until our bodies vibrated with the sound. They pressed against one another, urinating and defecating, their faces streaming with the fresh black stain of temporal gland secretions.

Who can know what goes on in the hearts and minds of elephants but the elephants themselves? What we had experienced was an intense greeting ceremony usually reserved only for family and bond group members who have been separated for a long time. I could only guess that I had been remembered and now, after all the time away, I had returned with other new, but familiar, smells: those of my brother, my mother, and my tiny daughter, held out to them in my arms.

Epilogue

JUNE 1993–SEPTEMBER 1995

A NUMBER OF SIGNIFICANT events have occurred in my life since the birth of Selengei. On June 2, 1993, when Selengei was three weeks old, Richard's aircraft developed engine trouble and he crashed in the foothills of the Aberdares. As a consequence he became a double amputee, losing both of his legs below the knee, and now walks with the aid of prostheses. In March of the following year, Richard resigned as Director of Kenya Wildlife Service following a protracted affair which, in itself, is a long and disturbing tale. A number of us, including myself, resigned in solidarity with him.

Following my resignation, I was able to devote myself to Selengei, and to the building of my house, which gave me deep satisfaction. The house I had initially built was, in fact, a small cottage, and now, with immeasurable help and advice once again from Richard, I supervised masons, carpenters, electricians, and plumbers in the building of the "big house." In early March of 1995, Selengei and I moved into our beautiful new home, and just as the first night in the cottage is etched in my memory, so, too, is our first night in the big house. I think it was the view of the night sky from the octagonal-shaped rooms and through large plate glass windows that was most extraordinary, and, after a few months here, I am "struck by a feeling," as Karen Blixen wrote in *Out of Africa,* "of having lived for a time up in the air."

Now, in a strange sequence of events, it seems that I may once again return to spend time with the Amboseli elephants. In late 1994, three of my study animals, Sleepy, RBG, and Sabore, were shot by licensed sport hunters when they crossed the international border into Tanzania. Cynthia became deeply involved in trying to stop any further hunting of the Amboseli bulls, and naturally, because of my scientific history with these individuals and my responsibility at KWS in setting cross-border elephant conservation policies, I, too, was deeply concerned at the turn of events. The shootings created an international furor, and in May of 1995 Tanzania agreed to a hunting moratorium until a thorough survey of the area, and its elephants, is undertaken. Together with Melly Reuling, a close friend and colleague from KWS, I put together a proposal to carry out a study, and in a marvelous twist of circumstance, the Tanzanian venture has renewed an old and important friendship with the companion of my youth, Paul Klingenstein.

As I write these words my chosen home, Kenya, is going through a particularly turbulent political period, and I have no clear sense of where it will lead. In May of 1995, Richard chose to join the Opposition, and because of our friendship I find myself feeling closer to the issues than I might have been otherwise. Sometimes I ask myself how I can continue to live in a place where there is such disregard for the law, and often such brutality. But, as the pink rays of the dawn touch the summit of *Esutuk,* across the valley, and the mist pours over the escarpment on either side of the house, I wonder how I could ever imagine leaving. "Can you see it, Mama?" asks a little voice beside me. "It is *so* beautiful!"

Index

African Elephant and Rhino
  Specialist Group (AERSG),
  195–98, 199–201, 207
African Elephant Conservation
  Coordinating Group
  (AECCG), 206–7
African Elephant Working Group
  (AEWG), 206, 210, 211
African Wildlife Foundation, 20,
  202, 213
Agamemnon (elephant), 45, 46,
  61, 74–76, 80
Alfred (elephant), 34, 80, 128
Amboseli National Park, 1–2, 7,
  8, 166–74, 238–39, 263–64
  drought in, 28–29, 30, 65
  Elephant Camp in, 22–27, 81–
  91

established as park, 40–41
filming in, 14, 201–2, 228, 273
Moss's elephant study in, 20–
  21, 23, 29–31, 213, 263; see
  also Moss, Cynthia
poaching and, 40, 263
Poole's childhood and, 9, 12
Poole's departure from, 239–40
Poole's move to, 20–21, 22,
  263
Poole's return visits to, 262
researchers vs. management in,
  182–91
Among the Elephants (Douglas-
  Hamilton and Douglas-
  Hamilton), 17, 49
Andelman, Rudolf, 125
Andelman, Sandy, 83, 125, 170

arap Moi, Daniel, 190, 209
Aristotle (elephant), 33, 34, 45,
    46, 61, 67, 167, 168

Bad Bull (elephant), 8, 45–46, 57,
    61, 63, 64, 71, 76–80, 103,
    105, 128, 166, 172
Baker, Liz and Neil, 230
Barnes, Richard, 60
Bashir, Abdul (Abdi), 246–47
Beach Ball (elephant), 34, 62
Big George (elephant), 273–74,
    275
Big Tuskless (elephant), 160, 192
birth, 92–94, 95–96, 192
Blixen, Karen, 257, 277
Bohlen, Buff, 202
bond groups (kin groups), 17,
    39–40, 41, 66, 67
  slaughter of elephants and, 208
  vocalizations and, 132
  see also clans
bones, 159–60, 161, 192, 193
brain, 140
  see also intelligence
bull elephants, see male elephants
Burrill, Anne, 196, 197–98
Burundi, 199
Bwana, Tahreni, 242, 247
Bygott, David, 60

calves, 17, 28–29, 37, 42, 65–66
  births of, 92–94, 95–96, 192
  lost, 145
  orphaned, 208
census, 17–18, 19–20
Chadwick, Doug, 228

Cheney, Dorothy, 83, 87, 133
Chloe (elephant), 273, 274
CITES, see Convention on
    International Trade in
    Endangered Species of Wild
    Fauna and Flora
clans, 40, 41–42, 66, 67
  Oltukai Orok, 84–85, 87
Cobb, Steve, 210, 231
communication, 137
  see also vocalizations
consciousness, see thinking and
    consciousness
conservation, 196, 200, 201, 203,
    206, 207, 232, 241, 246, 277
  see also population, elephant
conservation organizations, 202,
    206
  WWF, 202–3, 206, 210, 212
Convention on International
    Trade in Endangered Species
    of Wild Fauna and Flora
    (CITES), 198–200, 210, 237
  AEWG, 206, 210, 211
  Appendix I listing and, 201,
    211, 228–33
Croze, Harvey, 48
Cyclops (elephant), 34, 46, 49, 61
Cyn, see Jensen, Cynthia
Cynthia, see Moss, Cynthia

Danley, Tom, 132
David (elephant), 45, 46, 61
death, 92, 95, 97–99, 159–62,
    164–65, 192, 193
  orphaned calves and, 208
  see also killing of elephants

Dionysius (elephant), 34, 45, 46, 48, 61, 71, 74–76, 80, 103, 167, 168
Dobson, Andy, 240, 260
Douglas-Hamilton, Iain, 16–18, 19, 20, 39, 45, 48, 49, 77, 195, 206, 212, 215, 217, 228, 233, 245, 246
    AERSG and, 195, 196, 197–98, 200, 201
    in Tsavo elephant count, 204–5, 206
Douglas-Hamilton, Oria, 16–17, 19, 201, 205, 215, 233
drought, 28–29, 30, 65–66

East African Wildlife Society, 207–9
eating, 138–39, 140
Edwards, Steve, 201
Elephant Camp, 22–27, 81–91
Elephant Memories (Moss), 125
Elephant Specialist Group, 195
    elephant census by, 17–18, 19–20
Eltringham, Keith, 48, 58
empathy, 162–65
Evans, G. H., 61
extinction, 201, 211, 232
    see also population, elephant

families, 17, 28, 29, 30, 31, 38, 39, 40, 42, 67
    of "mystery females," 41
    naming of, 36
    slaughter of elephants and, 208
    vocalizations and, 132

    see also bond groups; clans
feeding, 138–39, 140
female elephants:
    birth and, 92–94, 95–96
    death of, 92, 97–99
    drought and, 28–29, 30, 65–66
    estrous, 28, 68, 73, 103, 104–6, 134, 137, 147, 167, 171, 208
    growth of, 68
    identification of, 29–30, 37, 42
    "mystery," 41
    orphaned calves and, 208
    social relations of, see social relations
    vocalizations of, see vocalizations
films, 14, 201–2, 228, 273
floppy runs, 154–55, 171
Fossey, Dian, 58
Francombe, Colin, 162–63

games, 150–58, 171
Gichangi, Steven, 243
Gombe, 182–83
Goodall, Jane, 58, 182
Green Penis (elephant), see Harvey
"Green Penis Syndrome," see musth, musth males
grief, 95
growth, 68–69
Guggenheim Foundation, 119, 120, 130

Hamilton, Patrick, 205, 206, 217, 223
Hanby, Jeanette, 60

Harcourt, Sandy, 60
Harvey (Green Penis) (elephant),
    44–45, 47, 57, 61
Hauser, Marc, 125
hearing, 120–21
Heffner, Rickye and Henry, 122
Hillman, Kes, 196
Hinde, Robert, 58, 59, 80, 81,
    101
Horner, Betty, 59
hunting, 18, 34, 40, 277
    see also poaching

Iain (elephant), 34, 57, 61, 62, 66,
    71, 80, 103, 166, 168
identification of elephants, 29–35,
    41–42
    naming and, 36–37, 42
intelligence, 140–41, 197
    ivory trade and, 197, 202
    see also thinking and
    consciousness
International Union for the
    Conservation of Nature
    (IUCN), 195, 201, 202
    elephant census conducted by,
    17–18, 19–20
ivory:
    elephants' reactions to, 160–61
    see also ivory trade; tusks
Ivory Record Book, 98, 188
Ivory Room, 98, 186
ivory trade, 34, 186, 187, 188,
    191, 196, 200, 201, 228–33
    ban on, 200, 201–3, 210, 211–
    12, 213, 228–33, 237, 250
    poaching in, see poaching

quota system in, 199, 200
species survival and, 197, 210,
    211; see also population,
    elephant
Ivory Trade Review Group
    (ITRG), 206, 210, 231

Japan, 232
Jarman, Peter, 18
Jensen, Cynthia (Cyn), 82–83, 86,
    96, 111, 113, 170–71
    Tonie and, 95, 96
Jezebel (elephant), 37–38, 40, 64,
    155, 161
Joshua (elephant), 155–57
Joyce (elephant), 37–38, 39, 40,
    155

Kasaine (elephant), 98, 99
Kasman, Lonnie, 72
Kenya, 9, 12, 13, 14, 15, 16, 18,
    184–85, 186, 190, 237–38,
    277
    elephant population in, 18,
    194, 198, 199
    ivory ban and, 229, 231, 232
    see also Amboseli National
    Park; Kenya Wildlife
    Service; Tsavo National
    Park; Wildlife Conservation
    and Management
    Department
Kenya Wildlife Service (KWS),
    243, 245, 246, 247, 250
    creation of, 238
    elephant surveys conducted by,
    245–46, 251–52

Leakey as director of WCMD
and, 231, 232, 276
Poole as Elephant Program
coordinator at, 238, 239,
240–42, 243, 247-48, 250–
56, 257, 262–64, 277
Poole's resignation from, 276
problem elephants and, 239,
240–41, 250, 251–56
Kilimanjaro, 1, 2, 4, 7, 23, 83,
88, 167, 170
Kilimanjaro elephants, 89–90
killing of elephants, 192–94
culling, 19, 256, 264
hunting, 18, 34, 40, 277
for ivory, see ivory trade;
poaching
"problem" elephants and, 239,
240–41, 250, 253, 254–55
kin groups, see bond groups
Kioko, Joe, 216, 222
Klingenstein, Paul, 17–18, 19–20,
50, 51, 277

Langbauer, Bill, 130
language, 142, 143, 144
Lasley, Bill, 72
Leakey, Mary, 268
Leakey, Meave, 258, 259, 260,
266, 270
Leakey, Richard, 1, 2–4, 208–9,
237, 238–39, 241, 242, 243,
244, 246, 247, 259–60, 262,
265–66, 270, 277
in airplane crash, 276
Poole's house and, 258, 259–
60, 266, 268, 276

as WCMD/KWS director, 231,
232, 276
Lee, Phyllis (Pili), 60, 81, 82, 100,
101, 102, 107, 113, 119, 124,
183, 208
book of, 182, 183
Lindsay, Keith, 77–78, 81–82,
100, 107, 113, 119, 151, 152,
154, 183, 207, 208

Maasai, 1, 2, 7, 9, 41, 89, 90,
165, 173, 183, 259, 265
elephants killed by, 193–94
farming by, 41, 42, 252
legends of, 111–12, 270–71
mores of, 109–11, 114–15
Selengei and, 270–72
see also Meloimyiet
McComb, Karen, 145
Mace, Ruth, 232
McKinder, Duncan, 101
McKnight, Barbara, 161
Maitumo, David, 77
Makulla, Simon, 163–65
Malawi, 9, 10, 12
male elephants:
"bull" or "retirement" areas
of, 66, 67, 69–70
dominance relations between,
68–69, 102–4, 147
female social relations vs., 17,
31, 67
female vocalizations vs., 131–32
"Green Penis Syndrome" in,
see musth, musth males
growth of, 68–69
identification of, 30–35, 42

male elephants *(continued)*
in musth, *see* musth, musth males
Manyara elephants, 18, 20, 39, 60, 125
Mapagoro, 182, 183
Martin, Rowen, 200
Masaku, *see* Sila, Masaku
Meloimyiet, 96, 97, 98, 108–16, 170–71
migration routes, 250, 264
Mikumi National Park, 213, 214, 230
Mlay, Costa, 230
Moss, Cynthia, 20–21, 22–24, 38, 41, 44, 47, 64, 65, 67, 81, 101, 104, 106, 112–13, 123, 124, 143, 147, 148, 170, 185–86, 208, 210, 211, 228, 263, 270, 273, 277
AECCG and, 206
AERSG and, 195, 196, 198, 200, 201
Bad Bull and, 45–46, 76, 77, 78–79
elephant identification by, 29–31, 33, 34, 42
Elephant Memories, 125
elephant naming and coding by, 36–37, 94
elephant society and, 39–40
grass surveys and, 151–53
ivory ban and, 201, 202, 203
musth ("Green Penis Syndrome") and, 44, 45, 47, 49, 58, 59
Poole's rift with, 212–13, 215
Portraits in the Wild, 20, 47

Muhoho, George, 208
musth, musth males ("Green Penis Syndrome"), 43–49, 57, 58–59, 60–64, 69–70, 102–6, 127–30, 137, 146–47, 166, 167, 168, 171, 172, 273, 274
in Asian elephants, 48, 60–61
association patterns in, 70–71, 102
believed not to occur in African elephants, 48, 58, 59
body size and, 68–69, 102, 103, 147
dangerous encounters with, 74–80
physical condition and, 69, 102, 103, 147, 168
Poole's design for study of, 57–64
Poole's dissertation on, 100–107
Poole's study of, 66–80, 92
Poole's thesis on, 59
slaughter of elephants and, 197, 208
temporal glands in, 44, 45, 48, 49, 62, 68, 168
testosterone in, 71–73
vocal and olfactory signals associated with, 119–20, 121–22, 123, 124, 133–34, 274
in younger males, 71, 104
Musth and Male-Male Competition in the African Elephant

(dissertation) (Poole), 100–
107
"Musth in the African Elephant"
(Moss and Poole), 59

Nairobi, 209, 239, 257–58, 262
naming of elephants, 36–37, 42
Nature, 59
Ngande, Peter, 125
Ngong Hills incident, 177–81,
183, 190, 206, 266
Njiraini, Norah, 127–30, 133,
158, 162, 190, 192, 255, 270,
271–72

Olindo, Perez, 206, 211, 214,
220, 229, 231, 232
Oloitipitip (elephant), 34, 61,
122, 172
Oltukai Orok clan, 84–85, 87
Ore, Arrghas, 215–19
Ore, Golo, 218, 219

Parker, Ian, 18, 19, 48
Payne, Katherine (Katy), 123–27,
130, 132, 134, 197
Pili, *see* Lee, Phyllis
poaching, 18, 19, 34, 40, 42, 44,
186, 187–88, 191, 193, 194,
195, 199, 200, 205–6, 208–9,
220, 229–30, 237, 243–48,
250, 263
Arrghas Ore's story of, 215–
19
methods used in, 217–18
Poole's study on, 208, 212–14,
220–27, 230, 231

in Tsavo, 205–6, 207, 208, 209,
213, 214, 216–17, 220, 222,
223
from vehicles, 205–6, 216–17
see also ivory trade
Polly (elephant), 97–99, 192
Poole, Amber Selengei, 270–72,
273, 274, 275, 276–77
Poole, Joyce:
brother of, 9, 10, 11, 13, 19,
50, 51, 52, 53, 58, 83, 258,
259, 260–62, 273–74
at Cambridge University, 58,
59–60, 82, 100–107, 112,
183
childhood of, 7–15
daughter of, 270–72, 273, 274,
275, 276–77
father of, 9, 10, 11, 12, 13, 14,
15, 17, 18, 19, 20, 22, 30,
31, 51–52, 53, 83
father's death and, 50–51, 52–
53, 57, 65, 260–2, 266, 273
house and land of, 258–59,
266, 267–69, 271, 276–77
love affairs of, 215, 224, 225,
231, 233, 258, 269; *see also*
Klingenstein, Paul;
Meloimyiet
mother of, 9, 10, 11, 12, 13,
14, 15, 19, 22, 50, 51, 52,
53, 57, 58, 83, 112, 227,
232, 273, 275
Ngong Hills incident and, 177–
81, 183, 190, 206, 266
pregnancies of, 231, 232–33,
269–70

Poole, Joyce (continued)
  returns to Africa, 16–21, 65, 81
  as single mother, 269–72
  sister of, 9, 10, 12, 13, 19, 52,
    53, 58, 83
  at Smith College, 45, 47, 48,
    49, 50, 59
population, elephant, 194, 199,
    213, 250–51, 256, 263–64
  in Amboseli, 263
  conservation and, 196, 200,
    201, 203, 206, 207, 232, 241,
    246, 277
  extinction and, 201, 211, 232
  human population and, 251; see
    also problem elephants
  in Kenya, 18, 194, 198, 199
  in Tanzania, 229–30
  see also ivory trade
Portraits in the Wild (Moss), 20, 47
problem elephants, 249–56
  killing of, 239, 240–41, 250,
    253, 254–55

Queen Elizabeth National Park,
    213, 214

RBG (elephant), 80, 277
Rees, A. F., 120
reproduction:
  drought and, 28–29, 65–66
  estrus, 28, 68, 73, 103, 104–6,
    134, 137, 147, 167, 171, 208
  Poole's study on poaching and,
    208, 212–14, 220–27, 230, 231
  see also musth, musth males
researchers, park management vs.,
    182–91

Reuling, Melly, 277
Rift Valley, 3, 258–59, 266, 267,
    268
Ruaha National Park, 60, 182,
    183
Rubenstein, Daniel, 119–20

Savage-Rumbaugh, Sue, 143
Sayialel, Soila, 130, 133, 158, 162
Scherlis, John, 60
Schindler, Paul, 202
Selengei, see Poole, Amber
    Selengei
self, sense of, 147, 160, 162, 165
sexual behavior, see reproduction
Seyfarth, Robert, 83, 87, 133
Sheldrick, Daphne, 160
Sikoi, 114
Sila, Masaku, 24, 25, 26, 82, 84,
    86, 87, 90–91, 125
Simaio, 114–15
Sleepy (elephant), 57, 80, 277
Slitear (elephant), 92, 93, 154
Slo (elephant), 154, 155
smell, sense of, 120–21, 137
  ivory and, 160–61
social relations, 17, 38–40, 66, 67,
    142, 202
  dominance, 68–69, 102–4, 147
  drought and, 30
  greeting ceremonies, 40, 137,
    275
  male vs. female, 17, 31, 67
  Poole's study on poaching and,
    208, 212–14, 220–27, 230,
    231
  vocalizations and, 124, 132; see
    also vocalizations

see also bond groups; clans;
   families
Somalia, 231
Species Survival Commission,
   195, 201
Stewart, Kelly, 60

Taabu (elephant), 25, 85, 87–88
Tallulah (elephant), 92–94
Tanzania, 199, 213, 214
   hunting in, 277
   ivory ban and, 229–30, 231, 232
teaching, 147
Teddy (elephant), 25, 85, 87–88,
   138
temporal glands, 44, 45, 48, 49,
   62, 68, 93, 97, 137, 168
testosterone, 71–73
thinking and consciousness, 141,
   142, 146–48, 149–50, 161, 165
   death and, 159–62, 164–65
   empathy and, 162–65
   imaginary enemies and, 148
   sense of self and, 147, 160, 162,
   165
   teaching and, 147
   vocalizations and, 142–46
Thomsen, Jorgen, 202, 206–7,
   210–12, 214, 228, 229, 230–31
Thompson, Bill, 228
Tonie (elephant), 84, 95–96, 192
tools, 139
   trunks and tusks as, 136–38,
   139, 140
TRAFFIC, 202, 228, 230
Trevor, Simon, 160, 205, 220,
   221, 223, 225, 226, 227
trunks, 136–38, 139, 140

Tsavo National Park:
   drought in, 28
   East African Wildlife Society
   conference in, 207–9
   elephant count in, 204–6, 207,
   216–17
   poaching in, 205–6, 207, 208,
   209, 213, 214, 216–17, 220,
   222, 223
Tuskless (elephant), 24, 25, 84,
   85, 87, 88, 150
tusks, 137, 138, 139, 161, 187,
   188, 224–25, 226–27
   see also ivory
Tyack, Barbara, 205, 215, 220,
   221, 257–58, 270

Uganda, 213, 214

Vee (elephant), 274, 275
Victoria (elephant), 274, 275
Virginia (elephant), 274–75
Vladimir (elephant), 8, 157–58, 162
vocalizations, 119, 122–27, 130–
   35, 137, 142–46, 197, 239
   attack rumbles, 144–45
   coalition rumbles, 145
   contact calls and answers, 124,
   145–46
   discussion rumbles, 143–44
   in females vs. males, 131–32
   "let's go" rumble, 124, 143,
   144, 158
   list of types of, 131
   lost calls, 124, 145

Wambua, Vincent, 125
Wanaina, Carole Mwai, 242

Wathoni, Pauline, 270, 271–72
WCMD, *see* Wildlife
    Conservation and
    Management Department
Weekly Review, 209
Western, David, 64, 69, 82, 198,
    207, 232, 264
Wickstrom, Dave, 132
Wildlife Conservation and
    Management Department
    (WCMD), 184, 186–91, 205,
    206, 207, 209, 214, 216–17

becomes Kenya Wildlife
    Service, 238
Leakey as director of, 231, 232,
    276
Wildlife News, 20
Wildlife Services, 19
World Bank, 238, 254
World Wildlife Fund (WWF),
    202–3, 206, 210, 212
Wrangham, Richard, 60, 64

Zeus (elephant), 43–45